VOLUME SEVENTEEN

RADIOACTIVITY IN THE ENVIRONMENT

TECHNOLOGICALLY ENHANCED NATURAL RADIATION

RADIOACTIVITY IN THE ENVIRONMENT

A companion series to the Journal of Environmental Radioactivity

Series Editor
M.S. Baxter
Ampfield House
Clachan Seil
Argyll, Scotland, UK

Volume 1: Plutonium in the Environment
(A. Kudo, Editor)

Volume 2: Interactions of Microorganism with Radionuclides
(F.R. Livens and M. Keith-Roach, Editors)

Volume 3: Radioactive Fallout after Nuclear Explosions and Accidents
(Yu.A. Izrael, Author)

Volume 4: Modelling Radioactivity in the Environment
(E.M. Scott, Editor)

Volume 5: Sedimentary Processes: Quantification Using Radionuclides
(J. Carroll and I. Lerche, Authors)

Volume 6: Marine Radioactivity
(H.D. Livingston, Editor)

Volume 7: The Natural Radiation Environment VII
(J.P. Mclaughlin, S.E. Simopoulos and F. Steinhäusler, Editors)

Volume 8: Radionuclides in the Environment
(P.P. Povinec and J.A. Sanchez-Cabeza, Editors)

Volume 9: Deep Geological Disposal of Radioactive Waste
(W.R. Alexander and L.E. Mckinley, Editors)

Volume 10: Radioactivity in the Terrestrial Environment
(G. Shaw, Editor)

Volume 11: Analysis of Environment Radionuclides
(P.P. Povinec, Editor)

Volume 12: Radioactive Aerosols
(C. Papastefanou, Author)

Volume 13: U-Th Series Nuclides in Aquatic Systems
(S. Krishnaswami and J. Kirk Cochran, Editors)

Volume 14: Remediation of Contaminated Environments
(G. Voigt and S. Fesenko, Editors)

Volume 15: Airborne Radioactive Contamination in Inhabited Areas
(K.G. Andersson, Editor)

Volume 16: Environmental Radionuclides: Tracers and Timers of Terrestrial Processes
(K. Froehlich, Editor)

Volume 17: Technologically Enhanced Natural Radiation
(A.S. Paschoa and F. Steinhäusler, Authors)

VOLUME SEVENTEEN

Radioactivity in the Environment

TECHNOLOGICALLY ENHANCED NATURAL RADIATION

Authors

A.S. Paschoa

Private Consultant, Rua Belisario Tavora 47, Rio de Janeiro, RJ 22245-070, Brazil

F. Steinhäusler

University of Salzburg, Division of Physics and Biophysics, Salzburg, Austria

ELSEVIER

Amsterdam • Boston • Heidelberg • London • New York • Oxford • Paris
San Diego • San Francisco • Singapore • Sydney • Tokyo

Elsevier
Radarweg 29, PO Box 211, 1000 AE Amsterdam, The Netherlands
Linacre House, Jordan Hill, Oxford OX2 8DP, UK

First edition 2010

British Library Cataloguing in Publication Data
A catalogue record for this book is available from the British Library

Library of Congress Cataloging-in-Publication Data
A catalog record for this book is available from the Library of Congress

ISBN: 978-0-08-101447-9
ISSN: 1569-4860

Printed and bound in Great Britain

10 11 12 13 14 10 9 8 7 6 5 4 3 2 1

Contents

Preface

When we decided back in January 2008 to jointly write a book on Technologically Enhanced Natural Radiation (TENR) we had no clear picture of the effort that would be involved in the project. Less than one month later, Murdoch Baxter, editor of the Elsevier's Radioactivity in the Environment Series sent Anselmo a stimulating message that convinced us of the importance of writing such a book. At that time, Anselmo was actively editing the Proceedings of the 8th International Symposium on the Natural Radiation Environment (NRE-VIII), which had been held in Búzios, Rio de Janeiro, Brazil, October 7–12, 2007; as a consequence there were superimposed time intervals in which he was finishing a project and starting a new one. The initial plan for the book was modified while writing it to avoid duplication with other books, which addressed the wide field of natural radiation. The book starts with a quasi-philosophical question of whether TENR is a Universal issue or not. The answer to this question is an unequivocal *yes*. TENR as a global issue is presented as a statement rather than a question. The evolution of global exposure rate estimates since 1959 is presented to show that such estimates have grown by a factor of two. The book explores current issues like depleted uranium (DU). The World Health Organization's (WHO's) views on the DU problem are presented, and the recommendations of the Institute of Medicine (IOM) of the National Academies (of the United States) on this subject are summarized. The longest chapter of the book deals with the intricacies of terrestrial, atmospheric, and aquatic natural radioactivity. Selected industries and their relationship to naturally occurring radioactive materials (NORM), technologically enhanced NORM (TENORM), and TENR raw material product, by-product, or residues are examined with regard to their potential to enhance radiation exposure to workers, consumers, and members of the general public. The issue of cosmic radiation, including its effects on airline crews, frequent flyer passengers, astronauts, and aircraft, satellites, and spacecraft equipment, is addressed in more than 30 pages, including a large number of tables and illustrative figures. The prehistory, history, and current aspects of metrology are presented in connection with their relation to NORM/TENORM/TENR. Environmental modeling regarding NORM/TENORM is briefly discussed, focusing on its objectives. We emphasize the need for sensitivity analysis of mathematical environmental models. Legal aspects of natural radiation are presented and discussed, including the current rather heterogeneous international approach. The need for harmonizing national and international regulatory frameworks to NORM/TENORM industries is pointed out, including guidance and practical implementation of existing concepts and principles. The potential of using natural radionuclides for making and using terrorist weapons is not ignored. Although for reasons of security, technical details are mostly absent, a discussion of this important current issue is presented, encompassing analytical aspects of the suitability of natural radionuclides already used for terror or homicide attempts. Illegal acquisition of natural radioactive materials, motivation, as well as modes that can be adopted for a terrorist attack with natural radionuclides, are discussed, as well as descriptions of what is usually called indistinctly *dirty bombs*. A risk assessment of radiological terrorism using natural radionuclides is presented as a guide to those involved directly or indirectly in impeding such terror attempts and attacks. A long list of references supports the statements, data, and overall information that appear in this book on TENR.

A book after being finished becomes a museum piece. However, people like Merril Eisenbud (1915–1997) and Konrad Bates Krauskopf (1910–2003) taught us that a book, like a museum, may be visited by those interested in learning from the past; teaching and undertaking research in the present; and developing new ideas, concepts, and research for the future. Thus, we feel a sense of achievement for making a little contribution to the vast book-museum that grows throughout the world every day.

We wish that this book will be helpful to students, scientists, and researchers in general, and to decision-making authorities an governmental and non-governmental organizations dealing with the NORM/TENORM/TENR issues. We hope also that officials in nuclear-, industrial-, aeronautical-, and space-agencies will find this book useful for making the critical assessment of the radiological impact of natural radiation vis-á-vis the radiological impact from man-made sources used in nuclear technology and nuclear medicine.

We owe gratitude to many colleagues in several countries who let us use their data, tables, and figures in this book. They are too many to be mentioned in this preface, but at least some of them are cited in the text. The TENR book could not have been written without their help.

In addition to the immense help of Murdoch Baxter, other members of the Elsevier staff helped us in different phases of the book project. We would like to mention at least those with whom we exchanged messages. They are the following: Linda Versteeg, Anita Koch, and Nicola Poser. More recently Mageswaran Babusivakumar made valuable suggestions and corrections to improve the quality of this book. All of the above-mentioned Elsevier staff members have helped at one time or another to overcome difficulties that looked at first glance unsurmountable. Our thanks are extensive to those Elsevier member staff who helped us *incognito* to put together this book.

Anselmo's wife, Alba, and his son, Claudio, provided permanent and necessary support to keep writing even when there were too many other commitments to attend to. We do have to confess that writing this book ends up being an enjoyable task.

Anselmo S. Paschoa
Saquarema, Rio de Janeiro, Brazil

Friedrich Steinhäusler
St. Gilgen, Salzburg, Austria
26 December 2009

CHAPTER 1

Introduction

1.1. Is TENR a Universal Issue?

Natural radiation (NR) is a phenomenon that pervades the whole known Universe. The term radiation is usually applied only to electromagnetic radiation which includes radio waves, microwaves, infrared rays (thermal radiation), visible light, ultraviolet, X-rays, and gamma rays – covering 19 orders of magnitude in terms of wavelength or frequency. The spectrum of electromagnetic radiation, also known as the Maxwell spectrum, is schematically illustrated in Figure 1.1.

The expression technologically enhanced natural radiation (TENR) encompasses all kinds of NRs which are – or were at any time in the past – enhanced by technological activities. Thus, one can ask the following germane question – Is TENR a Universal issue?

The concept of Universe as mentioned here extends from a radius of 5.5 light hours (or less) for the solar system to about 15 billion light years for the observable Universe. The existence of the cosmic microwave background radiation (CMBR) was predicted some 60 years ago by George Gamow (Gamow, 1948a; Gamow, 1948b). Gamow suggested that the Universe was a sphere of radius R made of matter and radiation, with R expanding proportionally with time t. At about the same time, Ralph Alpher and Robert Herman estimated the temperature of the CMBR to be between 5 and 28 K (Alpher and Herman, 1948). Not until 1965 was the CMBR detected experimentally as electromagnetic radiation (Penzias and Wilson, 1965). The current estimate of the CMBR temperature is about 2.725 K (http://www.eso.org/public/outreach/press-rel/pr-2008/pr-13-08.html – accessed on July 23, 2008).

The three governing laws of radiation in the Universe are usually summarized today as the following:

1. Planck's law (black body) – which relates the radiant energy emitted by a given celestial body (object) to the inverse of the fifth power of the wavelength of the radiation and the temperature of the radiant object;
2. Stefan–Boltzmann law – which relates the total energy emitted by an object to the fourth power of temperature; and
3. Wien's law – which establishes that the peak radiance decreases linearly as the temperature increases.

Radioactivity in the Environment, Volume 17
ISSN 1569-4860, DOI 10.1016/S1569-4860(09)01701-X

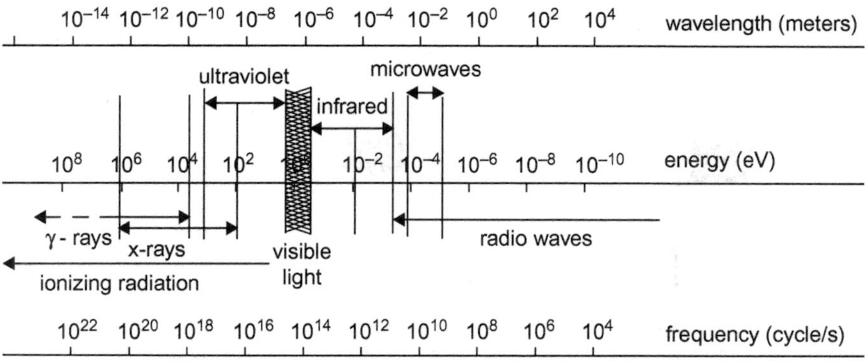

Figure 1.1 Electromagnetic spectrum or Maxwell spectrum.

When astrophysics was still considered a new scientific discipline, back in 1965, William A. Fowler was invited to give the fifth series of the Jayne Lectures at the American Philosophical Society in Philadelphia (Fowler, 1967). In his lectures, Fowler discussed the theoretical differences between the neutrinos and antineutrinos which appear in weak beta decay processes: for neutrinos, positron radioactivity, electron capture by positrons, and neutrino capture by neutrons; and for antineutrinos, electron radioactivity in the neutron decay process, positron capture by neutrons, and antineutrino capture by positrons.

The neutrino (and the antineutrino) had been postulated more than 30 years before Fowler's lecture in a letter written by Wolfgang Pauli to Lisa Meitner and Hans Geiger who were attending a meeting in Tübingen in December 1930 (Pauli, 1930). Pauli's letter preceded the discovery of the neutron by Chadwick by 2 years (Chadwick, 1932a, 1932b). The neutrino was finally detected by Frederick Reines and Clyde Cowan in 1956 (Reines and Cowan, 1956; Cowan et al., 1956).

The neutrino is a weak interacting particle with near-zero mass which travels with almost light speed. The very existence of the neutrino makes radiation a Universal issue.

Neutrinos are also produced on Earth either by natural radioactivity or in accelerators and nuclear reactors by the decays of natural radionuclides from the uranium isotope series (^{235}U and ^{238}U) and from the thorium (^{232}Th) series plus those from artificial radionuclides like plutonium and other transuranium isotopes. Neutrinos are also produced in the Sun and throughout the Universe in a myriad of nuclear reactions, including fusion.

There are three types of neutrinos ordered by increasing order of rest mass: electron-neutrino, muon-neutrino, and tau-neutrino. Some neutrinos detected on Earth are messengers from stars, including our Sun.

Neutrinos (i.e., geo-neutrinos) resulting from the decay chains of the ^{238}U, ^{235}U, and ^{232}Th plus the beta decay of ^{40}K are extant in the Earth's environment. The production of geo-neutrinos can be summarized by the series of simplified nuclear reactions as represented below:

$$^{238}U \rightarrow {}^{206}Pb + 8^{4}He + 6e^{-} + 6\bar{\nu}e$$
$$^{235}U \rightarrow {}^{207}Pb + 6^{4}He + 5e^{-} + 5\bar{\nu}e$$
$$^{232}Th \rightarrow {}^{208}Pb + 6^{4}He + 4e^{-} + 4\bar{\nu}e$$
$$^{40}K \rightarrow {}^{40}Ca + e^{-} + \bar{\nu}e$$

Several ongoing and future experiments are aimed at detecting geo-neutrinos (Bohem et al., 2001; SNO Collaboration, 2004; Fiorentini et al., 2005; de Meijer et al., 2006; Enamoto, 2006; Giammarchi, and Miramonti, 2006; Rubia, 2006; The Borexino Collaboration, 2006; Domogatsky et al., 2006; Learned, 2007; Raghavan, 2007; Neutrino Geoscience, 2008).

Neutrinos can also be used, however, in nuclear safeguards activities as a nonintrusive tool to check a reactor's activities (see, e.g., Bohem et al., 2001; Anjos et al., 2006; Bowden, 2008).

Figure 1.2 is an idealized graph of the rate of antineutrinos detected in (or thermal power of) a nuclear reactor as a function of time. The expected number of detected antineutrinos is of the order of a few hundred per day, depending on a number of parameters, which can be translated into the thermal power of the nuclear reactor.

Although one cannot say that humans can enhance the NR extant throughout the Universe to any significant degree, the Palo Verde, Double Chooz, and Angra dos Reis neutrino oscillations experiments show that nuclear reactors contribute a detectable number of electron–antineutrinos per fission of ^{235}U, ^{238}U, ^{239}Pu, and ^{241}Pu.

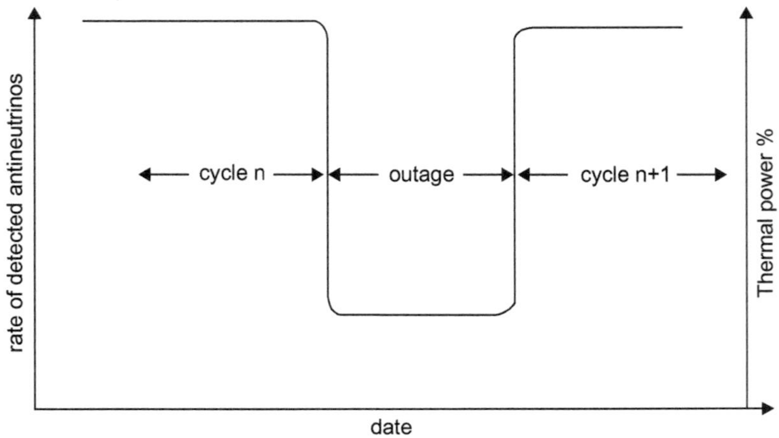

Figure 1.2 Idealized graph of the rate of antineutrinos detected in (or thermal power of) a nuclear reactor as a function of time.

Neutrinos are also produced artificially in nuclear detonations and particle accelerators (Zuber, 2004). Those neutrinos, like all other neutrino types mentioned above, are spread out all over but cannot be distinguished from those of natural origin.

Thus, one can say confidently that TENR is a Universal issue but, because the neutrinos interact very little with matter, the radiation doses from neutrinos do not need to be considered as far as radiation protection issues are concerned.

1.2. TENR – A Global Issue

The acronym TENR stands for technologically enhanced natural radiation, though the most commonly found definition in Internet dictionaries is "technologically enhanced natural radioactivity." The latter is a more restrictive definition derived from the wide use of the acronyms NORM (naturally occurring radioactive materials) and TENORM (technologically enhanced NORM).

As is well known, radioactive emissions encompass only the lower end of the Maxwell electromagnetic spectrum of radiation – as wavelength $10^{-15} > \lambda(\text{m}) > 10^{-7}$ or as frequency $10^4 < \nu$ (Hz) $< 10^{22}$ – while radiation *per se* includes the whole Maxwell spectrum represented schematically in Figure 1.1. As one can see in Figure 1.1, the spectrum of visible light encompasses just a small portion of the Maxwell spectrum, corresponding to a narrow window from 400 to 700 nm in the middle of the wavelength spectrum, while gamma rays, for example, have wavelengths of 10^{-10} m. The Maxwell spectrum ranges from less than 10^{-15} m to more than 10^4 m. This book will address TENR as technologically enhanced natural radiation in a broader sense, but without emphasizing ultraviolet, visible, infrared, microwaves, and radio waves. In other words, this book will deal essentially with natural ionizing radiation as far as it is enhanced technologically.

One can confidently state that there are only two types of NR: cosmic radiation and terrestrial radiation. Some authors, however, do prefer to add internal sources of NR in the human body as a third type of radiation.

Here, it is worth mentioning that cosmic radiation (or cosmic rays) is defined in a pedestrian way as high-energy particles of extraterrestrial origin. A more classical definition states that cosmic rays are nuclei (or ionized atoms) which can be from a single proton up to an iron nucleus and beyond. Protons constitute 90% and alpha particles 9% of all cosmic rays, consequently *all other* components are less than 1%. *All other* means stable and quasi-stable particles like neutrons, antiprotons (and maybe anti-nuclei), hard gamma rays (i.e., $\lambda < 10^{-14}$ m), electrons, positrons, neutrinos, antineutrinos, esoteric particles like weakly interacting massive particles (WIMPS) – which are hypothetical particles assumed to be a possible solution for the dark matter problem (see, e.g., Feng and Kumar, 2008) – magnetic monopoles, mini black holes, etc. These are the primary cosmic rays (PCRs) and are produced in stellar flares (like solar flares), supernovae events, and in other extraterrestrial galactic and extragalactic energetic sources.

The primary cosmic ray spectra at 10.6 GeV per nucleon used in calculations are composed of protons in the hydrogen flux (the unbound or free protons) plus 11 groups of primary nuclei, as shown in Table 1.1.

The interactions between the PCRs and the atmospheric and terrestrial nuclei produce secondary cosmic rays (SCRs). The SCRs are essentially all elementary particles and stable and unstable nuclei from which the most important are the following: nucleons, nuclei and nuclides, hard gamma rays, mesons ($\pi^{\pm}$, π^{0}, $K^{\pm}$, …, $D^{\pm}$, …), charged leptons ($e^{\pm}$, $\mu^{\pm}$, $\tau^{\pm}$), and neutrinos and antineutrinos (ν_e, ν_{μ}, ν_{τ}).

Concerns for the potential radiation doses to crew and passengers of high-altitude supersonic commercial air transport due to cosmic rays from galactic and solar events started in the early 1960s (Foelsche, 1961; Foelsche and Graul, 1962). However, the interest in the radiation doses due to cosmic rays had begun in the 1950s (UNSCEAR, 1956 and 1958, WHO, 1959). There are not so many differences between the cosmic ray doses received by high-altitude populations and those received by crew and passengers of commercial airplanes flying at altitudes varying from 3 to 12 km. The latter data are now available through a large number of publications.

One of the first important actions of UNSCEAR was to carry out investigations on the biological effects of ionizing radiation (UNSCEAR, 1956, 1958). It was then accepted that the probability of mutation per *locus*

Table 1.1 Components of the primary cosmic ray spectrum at 10.6 GeV per nucleon (O'Brien, 2005).

Z	Element	Relative abundance	%
1	H	750	95.233
2	He	34	4.317
3–5	Li–Be	0.4	0.051
6–8	C–O	2.20	0.279
9–10	F–Ne	0.3	0.038
11–12	Na–Mg	0.22	0.028
13–14	Al–Si	0.19	0.024
15–16	P–S	0.03	0.004
17–18	Cl–Ar	0.01	0.001
19–20	K–Ca	0.02	0.003
21–25	Sc–Mn	0.05	0.006
26–28	Fe–Ni	0.12	0.015

per unit dose was of the order of 1×10^{-7}. As a consequence, it was also known that a rather large number of observations needed to be made to resolve the "*signal to noise ratio*" to obtain statistically significant results concerning the *dose response relationship*. Not long after the 1958 UNSCEAR report was published, a World Health Organization (WHO) Expert Committee also produced a report which expressed concern about the genetic effects which might be produced in humans due to the increasing use of ionizing radiation in medicine, science, and industry (WHO, 1959). It was then expected that the study of human populations exposed to relatively large amounts of background radiation (i.e., of the order of 1 rem y^{-1} – or 10 mSv y^{-1} in today's units) would bring untapped information on radiation-induced mutations and their fate.

As mentioned by Paschoa (2000a, 2000b), some of the obvious populations to be studied, in accordance with the WHO (1959) committee recommendations, were those living in high-altitude areas; for example, Cerro de Pasco, Peru, 4.3×10^3 m, latitude 10°S; the Himalayan area (Lhasa), 3.7×10^3 m, latitude 30°N; La Paz, Bolivia, 3.6×10^3 m, latitude 16°S; Quito, Ecuador, 2.9×10^3 m, latitude 0°; Bogota, Colombia, 2.6×10^3 m, latitude 4°N. However, areas with high natural radioactive background were also, in some cases, worth investigating. Among those, a WHO Committee mentioned the following (WHO, 1959): part of the Kerala State, in India, and adjoining area in Madras State; the monazite areas in the Brazilian States of Espírito Santo and Rio de Janeiro; the mineralized volcanic intrusives in the Brazilian States of Minas Gerais and Goiaz; the primitive granitic, schistous, and sandstone areas of France with slight elevation of NR; and some uninhabited areas of the Belgian Congo.

At that time, the doses from external and internal irradiation from natural sources under the usual conditions at sea level were believed to be as shown in Table 1.2 (WHO, 1959). The temporal evolution of concepts and the improvements of knowledge and measurement techniques resulted in new information on the doses from external and internal irradiation. The 1959 dose estimates did not have data on the inhalation exposure of either radon or thoron, though it had estimates for ^{14}C, because of the importance which was then given to the introduction of this radionuclide into the biosphere due to the nuclear weapon tests in the atmosphere (Suess, 1953; Revelle and Suess, 1957; Arnold and Anderson, 1957; Young et al., 1965). Ten years before the WHO Expert Committee met, ^{14}C had started being used as a reliable dating tool based on the ratio $^{14}C/^{12}C \approx 1.3 \times 10^{-12}$ found both in living organisms and in the atmosphere (Arnold and Libby, 1949; Libby, 1955). However, there was the suspicion after the atmospheric nuclear weapon tests that the carbon isotopic ratio would change to an extent that would affect the reliability of the method (see, e.g., Krane, 1987).

Table 1.2 Exposure rates from external and internal irradiation to gonads and other soft tissues, as known in 1959.

	Exposure rates	
	mrem y^{-1}	μSv y^{-1}
External irradiation		
Cosmic rays	28	280
Gamma rays (outdoors)	47	470
Internal irradiation		
^{40}K	19	190
^{14}C	1.6	16
^{226}Ra	?	?
Total (from all sources)	95	950

Source: Adapted from WHO (1959) – see also Paschoa (2000a, 2000b).

Table 1.3 compares as much as possible the dose estimates made by WHO (1959) with those made by UNSCEAR (2000). It is interesting to note, by observing the last column of Table 1.3, that though the dose estimates for cosmic radiation increased and ingestion exposure due to terrestrial radionuclides increased by only about 30% each, and the dose estimates from terrestrial external irradiation did not change significantly, the total dose estimates increased by about 150%. Radon (^{222}Rn) inhalation which was not accounted for in the WHO estimates in 1959 became by far the most important individual component of the dose estimates made by UNSCEAR in 2000.

The areas with high levels of natural radioactivity and/or high radon or thoron concentrations are likely to become TENR areas because of mineral exploitation and later industrial processing of NORM material, as has been pointed out in the Brazilian case (Paschoa, 2002). The WHO Committee (WHO, 1959) had suggested that populations exposed to annual doses of about 10 mSv (i.e., ≈4 times higher than the average dose to the world population), as per the UNSCEAR (2000) estimate, would bring untapped information on radiation-induced mutations and their effects. Much has been achieved as far as the dose assessment of elevated NR areas is concerned (see, e.g., Eisenbud, 1982; Sohabi and Esmaili, 2002). Occupational annual dose equivalents might have reached levels as high as 55 mSv from thoron inhalation prior to the improvement of industrial hygiene in a Brazilian monazite industry (Paschoa and Pohl-Rülling, 2005). Such high annual doses are more than 500 times the current estimate for the world average annual thoron inhalation dose, or approximately 20 times the world average dose from all sources.

Table 1.3 Exposure rates from external and internal irradiation as estimated by UNSCEAR (2000) plus the ratio WHO (1959)/UNSCEAR (2000) of those estimates.

	Exposure rates		
	WHO (1959) $\mu Sv\, y^{-1}$	UNSCEAR (2000) $\mu Sv\, y^{-1}$ [a]	WHO (1959), UNSCEAR (2000)
External irradiation			
Cosmic radiation	280		
Directly ionizing plus photon component		280	
Neutron component		100	
Cosmogenic radionuclides		10	
Subtotal (cosmic plus cosmogenic radionuclides)	280	390	0.72
External terrestrial irradiation			
Gamma rays (outdoors)	470	70	
Gamma rays (indoors)		410	
Subtotal (external terrestrial irradiation)	470	480	0.98
Internal irradiation			
Inhalation exposure			
Uranium and thorium series		6	
Radon (^{222}Rn)		1.15×10^3	
Thoron (^{220}Rn)		100	
Subtotal (inhalation exposure)	?	1.26×10^3	?
Ingestion exposure			
^{40}K	190	170	
Uranium and thorium series		120	
^{14}C	16		
^{226}Ra	?		
Subtotal (ingestion exposure)	206	290	0.71
Total (from all sources)	950	2.42×10^3	0.39

[a] Average.

The importance of evaluating occupational and public doses from TENR areas becomes clear. This fact has been stressed recently in the NORM V Symposium (Wymer, 2008).

1.3. Overview

Human activities can modify NORM into TENORM, resulting in enhanced concentration of NORM in a product, by-product, or residual material. Thereby individuals can receive increased radiation exposure due to TENR either as workers or consumers. Such elevated radiation exposure can also result from extreme environmental exposure situations outside of direct control, such as high-altitude flights and activities in space due to cosmic radiation. The following section provides an overview of these main topic areas, describing schematically the individual production steps, end products, and associated emissions and wastes (TENR compartments marked in grey) (IAEA, 2003a, 2003b).

1.3.1. From NORM to TENORM due to human activities

Environmental conditions can change as a result of several industrial processes, resulting in an increase in the concentrations of natural radionuclides in the material produced as the end product or waste material; examples of such industrial processes are shown schematically in Figure 1.3.

Each of these processes which appear in the top layer boxes of Figure 1.3 can result in an increased exposure to TENORM, for example:

(a) the extraction of water from subsurface aquifers can result in the precipitation of scales, containing elevated levels of radium (^{226}Ra);

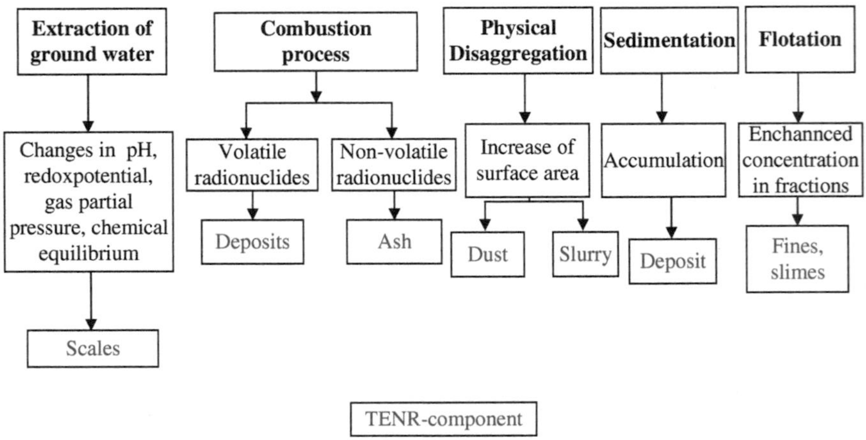

Figure 1.3 Industrial activities leading to increased concentrations of natural radionuclides. TENR compartments marked in grey.

(b) processes burning fossil fuel and hydrocarbons can lead to atmospheric emission of volatilized lead and polonium and increased concentration of radium in ash;
(c) NORM undergoing physical disaggregation may undergo enhanced dispersion of natural radionuclides in the environment;
(d) NORM subjected to sedimentation may further increase its concentration of natural radionuclides; and
(e) several fractionation processes can enhance NORM levels further in the end product, such as in slimes or fines.

1.3.2. Hydrocarbons

Hydrocarbons are usually extracted from reservoirs located in sediments. The use of *formation water*, a mixture of oil–gas–water in contact with the reservoir exploited, can lead to an increased dissolution of radium (^{226}Ra) in comparison to uranium (^{238}U) and thorium (^{232}Th). These radionuclides, together with their decay products (^{222}Rn, ^{220}Rn, and their decay products), will contaminate the surfaces of equipment, pipes and tanks, and potentially the environment (Figure 1.4).

Several subsequent stages in the production of different fuels (LNG, ethane, propane, etc.) will also be accompanied by the emission of NORM into the environment. Workers in charge of maintenance can receive occupational radiation doses from a wide variety of exposure scenarios, ranging from the cleaning of tanks to handling pipes and valves coated with NORM.

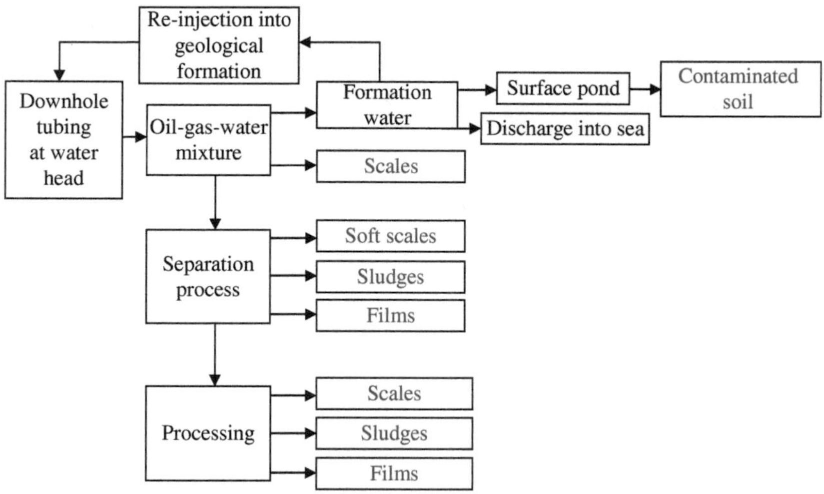

Figure 1.4 NORM in the oil and gas industries.

Figure 1.5 illustrates that a particular issue to be addressed can be the exposure to radon and its decay products and emission during oil and gas extraction and front-end reduction process.

1.3.3. Coal

Coal will be in increasing demand in the next decades at least – besides oil and gas – as one of the main sources of primary energy worldwide. Fossil fuel represents NORM (mainly radium and its decay products) so that increasing extraction of this fossil fuel from underground mines and surface mines will result in TENORM issues. These issues cover a wide range, from occupational radiation exposure of miners, radioecological challenges due to atmospheric emissions of radium and its decay products from coal-fired power plants, to managing the large amounts of waste (waste rock, waste water, ash, and slag) due to mining and combustion processes. Figure 1.6 shows schematically the different stages of the coal fuel cycle, from mining to combustion as fuel in a coal-fired power plant.

TENORM arises as an elevated radon and decay product concentration due to mining, to radium-containing liquid effluents, as well as in sludge, sediments, tailings, waste rock, and gypsum. In view of the large amounts of residues, complex waste management schemes are required for an environmentally sound and cost-effective disposal.

1.3.4. Geothermal energy

High temperature fluids, heated in geothermal reservoirs below the Earth's surface, are increasingly utilized for energy production in geothermal power plants, for geothermal heat pumps or used in direct heat applications. In

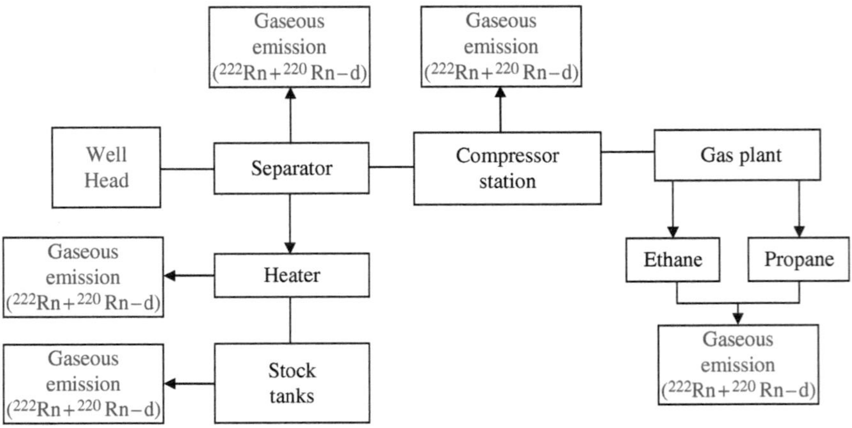

Figure 1.5 Radon, thoron (^{222}Rn, ^{220}Rn), and decay products and emissions during oil and gas extraction and front-end reduction process.

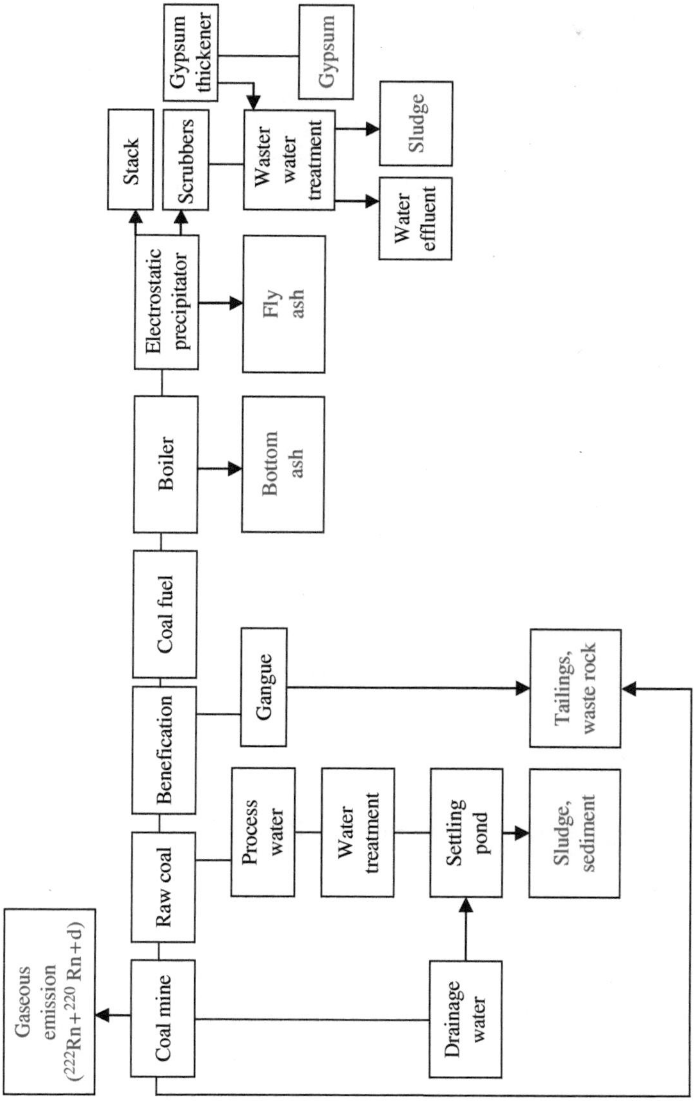

Figure 1.6 Coal fuel cycle.

either case, boreholes drilled into the underground reservoir are used to utilize the heated fluid above ground in heat exchangers and steam turbines, as indicated in Figure 1.7.

1.3.5. Water treatment

Most water supplies used for drinking water or industrial process water require treatment in order to remove impurities, such as chemicals and metals, in some cases also NORM (e.g., ^{226}Ra). Also discharges by various industries necessitate treatment of the wastewater (e.g., mining industry). This treatment is carried out in waterworks, as illustrated in Figure 1.8.

The NORM in such water reflects the local geology, typically containing uranium (^{238}U, ^{234}U), radium (^{226}Ra, ^{228}Ra), and radon

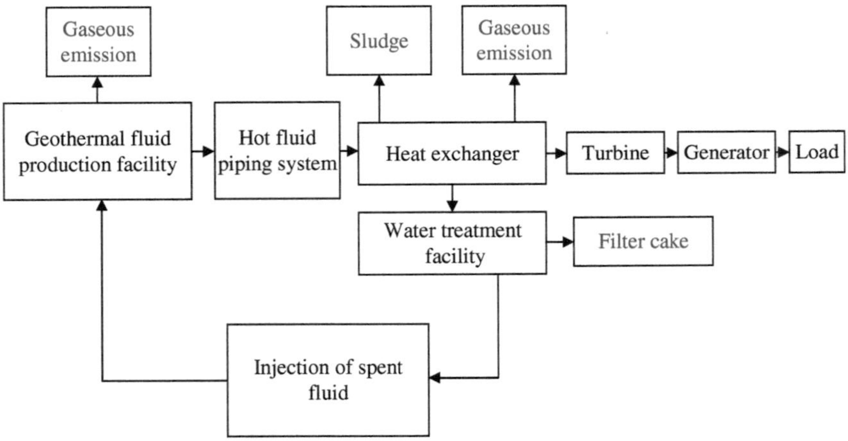

Figure 1.7 Geothermal power usage.

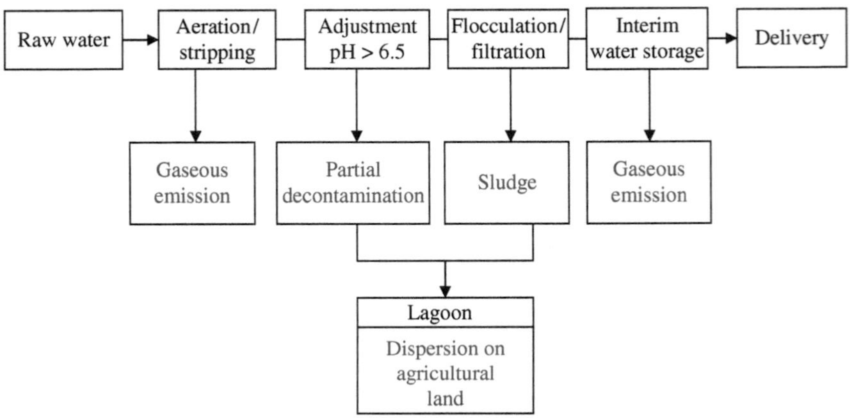

Figure 1.8 Water treatment process.

(^{222}Rn). Water treatment can result in a significant amount of TENORM due to the large volume of water treated and the increased concentration of natural radionuclides in filtration equipment, such as spent ion exchange resins, spent granular activated carbon, sand filter elements, together with contaminated back flush water, filter sludge, and lime sludge.

1.3.6. Metal mining

Metals are mined through underground mines, open pit mines, and solution mining. Each of these techniques generates large amounts of solid and liquid wastes. Every stage of the industrial process involves NORM, from the extraction of the ore, to its physical or chemical processing, and finally the smelting of the ore in order to extract and purify the metal. Since the process feed ore contains NORM, the residues, tailings, and slag from each stage can contain elevated levels of natural radionuclides, as illustrated in Figure 1.9.

1.3.6.1. Aluminum production

The main source for aluminum is bauxite ore, which contains approximately 50% $Al_2O_32H_2O$. Bauxite ore contains radionuclides from the uranium and/or thorium series. The production process is associated with large amounts of residues, posing economic and ecological difficulties (typically between 1.1 and 6.2 ton red mud per ton alumina produced). Figure 1.10 provides an overview of the main sources leading to TENR.

The major such waste component is *red mud*, an alkaline slurry, which is deposited in heaps or depressions. Radon and thoron emanation, as well as airborne dust, are potential sources of NORM exposure (USGS, 2008).

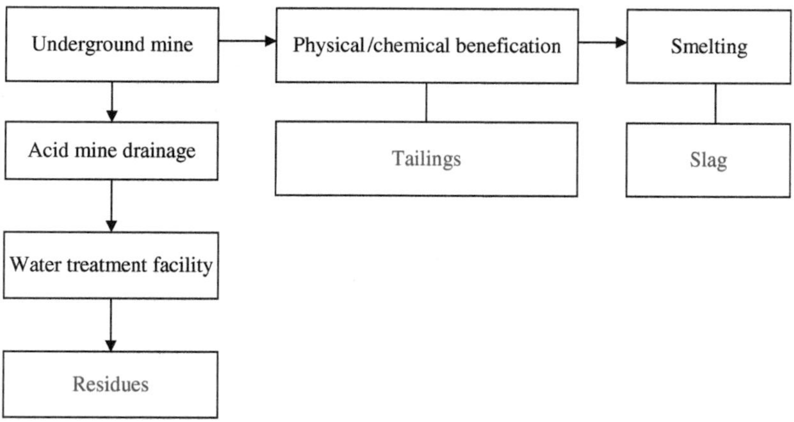

Figure 1.9 Metal mining and processing.

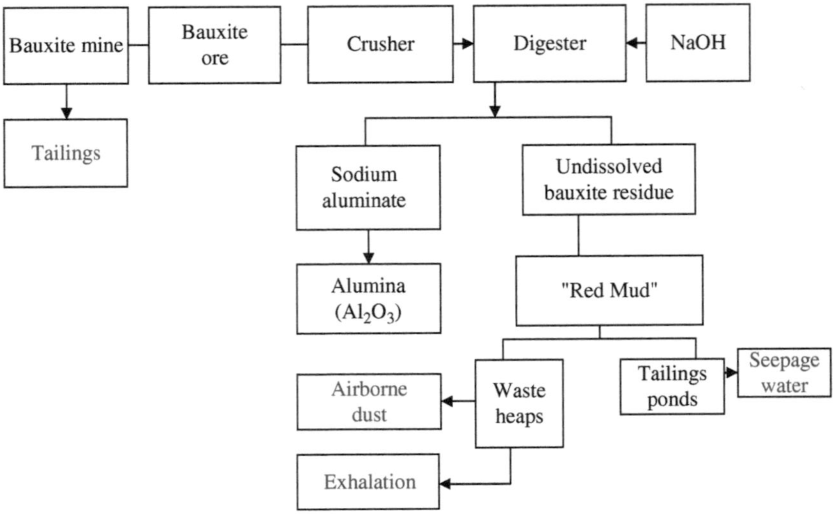

Figure 1.10 Aluminum production.

1.3.6.2. Iron production

Iron ores, especially in a state of enhanced weathering, scavenge natural radionuclides. In addition, major components used for iron and steel making also contain uranium. This can lead to TENORM in off-gas cleaning dust. Also, slag and off-gas cleaning sludge are likely to contain uranium decay products at increased levels. Figure 1.11 provides an overview of the different compartments contributing to TENR.

1.3.6.3. Gold production

Worldwide gold is used as jewelry, support for currencies, in electronic appliances and optical instruments (e.g., lasers), and in the construction and space industries (e.g., shielding from solar radiation). Global gold production is about 2,500 ton y^{-1} (USGS, 2008). In view of the fact that ore mined for gold contains only about 5 g ton^{-1}, global mine tailings amount to approximately 10^9 ton annually. In Figure 1.12, mining and subsequent metal production are shown schematically. The main components of the gold-production-related TENR are seepage and surface runoff in the tailings area, radon exhalation from tailings, waste rock crushed and used as construction material, as well as contaminated materials (e.g., wood from underground mining, bulk wastes) and scales formed during operations.

1.3.7. Industrial processes

Raw materials used in a variety of industrial processes containing NORM can result in TENR. These processes range from the mining and use of different mineral sands and phosphates, to the production of bricks from

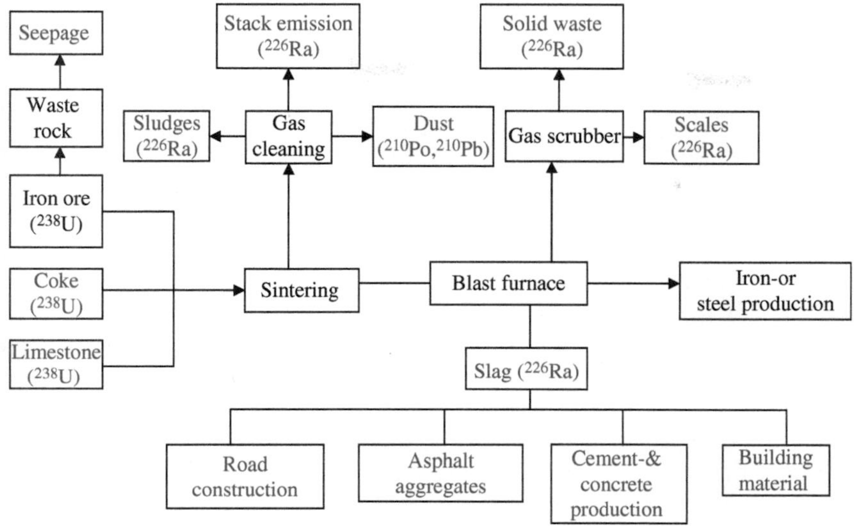

Figure 1.11 Iron and steel production.

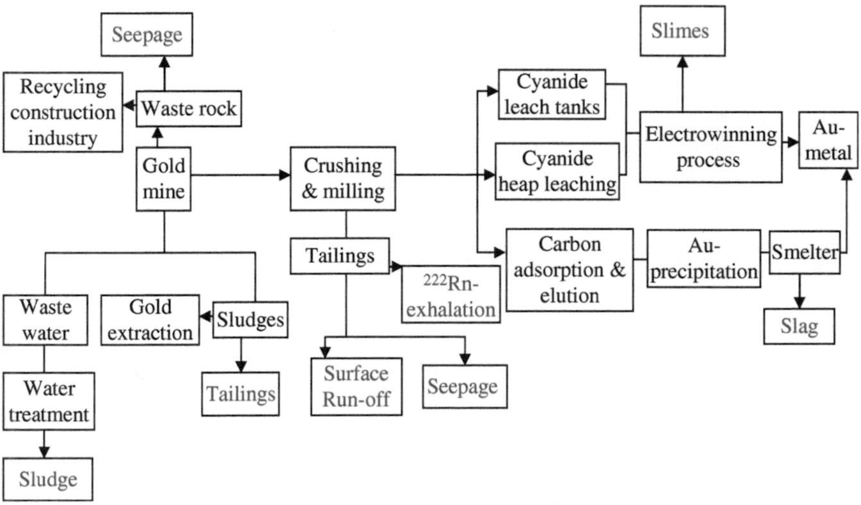

Figure 1.12 Gold production.

certain raw or recycled materials, to recycling of NORM-containing wastes resulting from decommissioned dwellings.

1.3.7.1. Mineral sands

About 5.5 million ton (Mt) of heavy minerals (density $>2.8\,g\,cm^{-3}$), such as monazite, ilmenite, rutile, zircon, yttrium, casserite, bastnasite, and xenotime, are mined annually worldwide; USGS, 2003). They serve as raw materials for the extraction of valuable elements, for example, titanium, zirconium, thorium, and rare earths. Such materials find widespread industrial uses for alloys, polishes, catalysts, ceramics, foundries and refractory manufacture, and glass production. In Figure 1.13, the main steps in separating mineral ore into its different products and the resulting waste streams are shown schematically.

1.3.7.2. Fertilizer production

Phosphorite is mined in about 30 countries almost exclusively in open pit mines. This phosphate rock is subsequently processed into the intermediates phosphoric acid and the element phosphorus. Global production amounts to about 49.8 Mt of P_2O_5 annually (Jasinski, 2007). The intermediates represent the basis for a range of important products in the agriculture and chemical industries (e.g., fertilizer, food additives, pesticides, detergents). In Figure 1.14, the mining and processing of phosphate ore are shown.

1.3.7.3. Building material

Natural material (sand, gravel, rock, clay, marble), recycled waste material (coal ash, phosphogypsum, furnace slag, slimes and fines from aggregate plants), or finished products (cement, tiles, gypsum plasterboard) can all contain largely differing values of NORM and thus residents living or employees working in dwellings with NORM or TENORM used in their

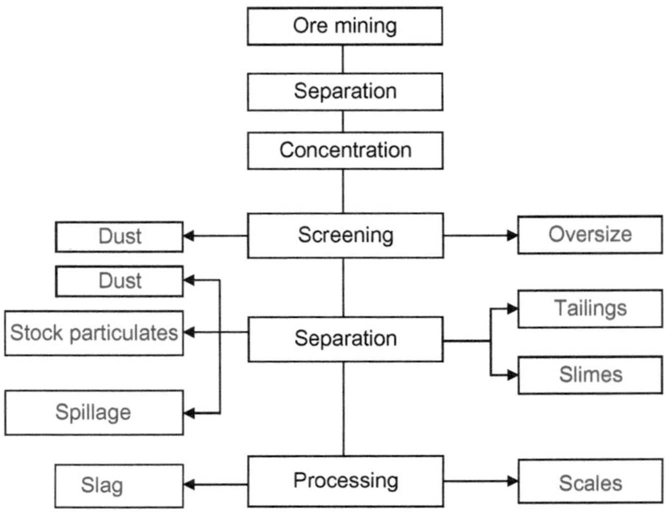

Figure 1.13 Heavy mineral ore separation scheme.

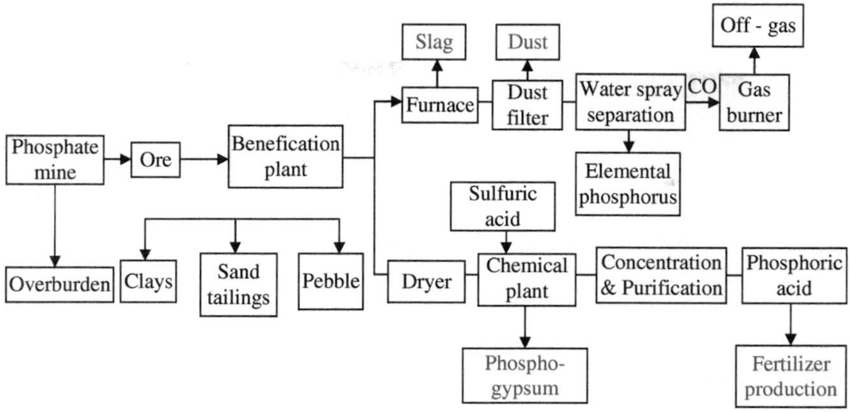

Figure 1.14 Phosphate ore mining and processing.

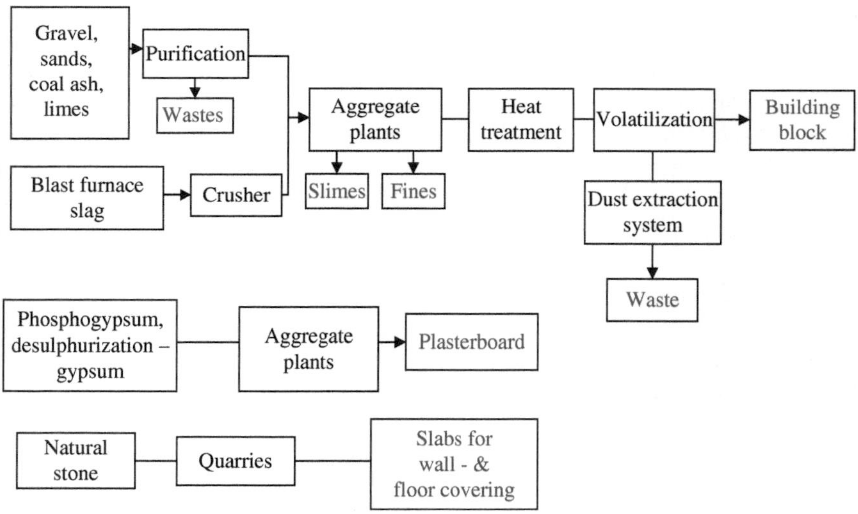

Figure 1.15 Processing of building materials.

construction may be exposed to a TENR environment for prolonged periods. The TENORM components due to processing of such material are described in Figure 1.15.

CHAPTER 2

Depleted Uranium (DU) as TENR

2.1. A Brief History of DU

Natural uranium (NU) is composed approximately of 99.27% ^{238}U, 0.72% ^{235}U, and 0.0055% ^{234}U. A by-product of the production of low enriched uranium (LEU) and highly enriched uranium (HEU) – that is, uranium enriched in ^{235}U to low or high percentages – used for fission in nuclear reactors and nuclear weapons, respectively, is called depleted uranium (DU). Once the ^{235}U-enriched fraction obtained in the enrichment process is removed, the typical DU remnants comprise 99.8% ^{238}U, 0.2% ^{235}U, and 0.001% ^{234}U. This means that DU contains less than a third of the ^{235}U than NU. During the Manhattan Project, the term "depletalloy" was used to refer to an alloy in which the fissile ^{235}U had been reduced (Raabe, 2002). In fact, the term depletalloy was derived from other terms used in the Manhattan Project like tuballoy (NU) and oralloy (enriched uranium). Other code names like Q-metal and D-38 were also used in the past. Stocks of DU are usually found as UF_6, U_3O_8, or metal. Here, it is worth mentioning that, on August 19, 1943, in Quebec Citadel overlooking the Plains of Abraham, Franklyn Delano Roosevelt and Winston Churchill signed the "Articles of Agreement Governing collaboration between the authorities of the U.S.A. and the U.K. in the matter of Tube Alloys" which became better known as the Quebec Agreement that first formalized the nuclear agreement between these two countries during World War II (WWII) (Hewlett and Anderson, 1962). In the context of the Quebec Agreement, the expression "Tube Alloys" was the code name for the British uranium program. At the end of WWII, there were attempts to use metallic uranium in armor-piercing ammunition (Gaca et al., 2005).

The density of metallic DU is 19.0 $g\,cm^{-3}$. This high density makes DU appropriate for use in several types of ammunition. DU is used, for example, to pierce armor plating, in cruise missiles, and in the armor of tanks. In 1968, a patent requirement was filed by Brevets, Aero-Mecaniques S.A. (Geneva, CH), for a long-burning pyrotechnic material containing DU for spotting rifle projectiles (US Patent, 1971).

DU-bearing projectiles started being widely used in the Desert Storm War (i.e., Gulf War, 1990–1991). Later the "Gulf War Syndrome" called the public and governmental attention to the potential consequences of

Radioactivity in the Environment, Volume 17
ISSN 1569-4860, DOI 10.1016/S1569-4860(09)01702-1

using DU in ammunition (Department of Veterans Affairs, 1995; Eisenbud and Gesell, 1997). DU has been used in antitank weaponry, and in tank shielding against regular weaponry.

As a matter of fact, the use of DU by the military in the Gulf War, and in the conflict in the Balkans, has caused considerable concern amongst the general public. NATO Member States reported the incidence of health effects and led to speculation on their potential link to an increased exposure by soldiers to DU in the line of duty (Steinhäusler and Paschoa, 2005). Moreover, claims have been made about the possible threat to the health of residents in DU-affected areas. The United Nations Environment Programme (UNEP) sponsored a DU postconflict environmental assessment in Kosovo, Serbia, and Montenegro (UNEP, 2001, 2002).

During the first surveys made by UNEP in Kosovo, only a NaI(Tl) detector with an alarm signal capable of being heard above the traffic noise and the wind could be used (UNEP, 2002). The UNEP surveys were carried out to locate potential DU-contaminated sites. Ge(Li) detectors could not be used at the time because it was not possible to transport the liquid nitrogen that was essential to cool them before applying bias voltage. However, after the publication of the surveys in Kosovo, the Yugoslavian authorities invited a UNEP team to carry out DU studies in Serbia and Montenegro (UNEP, 2002). This time a large number of soil samples were collected to be measured subsequently. The NU levels of the soils at the sites of collection ranged from 1.0 to 9.5 mg U kg^{-1} soil (UNEP, 2002). Taking into account the 0.7% ^{235}U abundance in NU, DU was defined by the UNEP team as having ^{235}U abundance in soil samples equal to or less than 0.35% (UNEP, 2002). From the soil samples collected in Serbia and Montenegro, it was determined that approximately 42% of these samples presented more than 10% DU. Here, one must bear in mind that farmers working in DU-contaminated soils, as well as children playing in such soils, might sometimes ingest small amounts of DU (IAEA, 2003a, 2003b; McLaughlin, 2005; Bikit et al., 2005).

Nowadays, one should be well prepared to evaluate the potential health and environmental impacts associated with the possibility of future military use of DU. As a matter of fact, the issue of DU recently led several investigators to review gamma-spectrometric techniques to discriminate between NU extant in the environment and DU in contaminated areas (Bikit et al., 2001; UNEP, 2002; IAEA, 2003a, 2003b; McLaughlin, 2005; Bikit et al., 2005).

Advantage has been taken at the beginning of the 21st century of low energy gamma-counting systems, put together originally for NORM and TENORM measurements, to develop simple gamma-spectrometric methods to measure DU (Paschoa, 2001; Paschoa et al., 2003; Bikit et al., 2005).

In the 1970s, when the use of lithium-drifted germanium detector [Ge (Li)] was still a novelty, few investigators attempted to measure ^{238}U through the 63.3 keV transition line from daughter ^{234}Th. This was so until the isotopic ratios $^{235}U/^{238}U$ (reported originally as $^{238}U/^{235}U$) were successfully determined by gamma spectrometry in soils with NU content, and contaminated with DU, and then compared with the isotopic ratios obtained by mass spectrometry (Coles et al., 1974). Soils with NU content were obtained at the Livermore California Valley and at Yosemite National Park, while those soils with DU contamination were collected at a high explosive test area (Coles et al., 1974). Table 2.1 summarizes the $^{235}U/^{238}U$ isotopic ratios in natural and contaminated soils estimated from the published data (Coles et al., 1974; Paschoa et al., 2003).

The soil characterized in Table 2.1 as having a low uranium level had approximately 2 ppm, and the ^{238}U content was inferred from the ^{226}Ra plus daughters and from the 63.3 keV gamma-ray line. By doing so, the gamma-counting statistical errors were large, as reflected in the uranium isotopic ratio, which is a function of the uranium content of the soil. This can be seen by observing in Table 2.1 the results obtained by gamma spectrometry for low and high uranium levels with the "natural" isotopic ratio. In addition, in the samples measured, ^{238}U was not in equilibrium with its long-lived daughters ^{234}Th and ^{226}Ra. As a consequence, the only ^{226}Ra-based U ratio measured, considered to be "natural," was that of the soil with high uranium content, that is, 0.045. This isotopic ratio is consistent with the results obtained by mass spectrometry for soils with low and high uranium levels with "natural" $^{235}U/^{238}U$ isotopic ratio. The agreement obtained for the results of the measurements of DU-contaminated soil, made either by gamma or mass spectrometry, indicates that the lower the ^{235}U content in the soil sample the better the agreement between the results obtained by the

Table 2.1 Isotopic ratio ($^{235}U/^{238}U$) in soils with NU and contaminated with DU as measured by gamma and mass spectrometries.

	$^{235}U/^{238}U$	
	Gamma spectrometry	Mass spectrometry
	0.027 ± 0.017	0.045 ± 0.001
	0.036 ± 0.016	0.047 ± 0.001
	0.050 ± 0.024	0.046 ± 0.001
	0.032 ± 0.014	0.046 ± 0.001
High uranium level with "natural" isotopic ratio	0.045 ± 0.003	0.046 ± 0.001
	0.012 ± 0.001	0.012 ± 0.001
	0.012 ± 0.001	0.013 ± 0.001
	0.014 ± 0.002	0.014 ± 0.001
	0.015 ± 0.003	0.016 ± 0.001
	0.013 ± 0.002	0.016 ± 0.001
	0.013 ± 0.001	0.013 ± 0.001

Source: Adapted from Coles et al. (1974) and Paschoa et al. (2003).

two techniques. One possible explanation for this effect may lie in the fact that the quantitative determination of ^{235}U was based on the 185.72 keV gamma ray, resulting from the alpha decay of ^{235}U into ^{231}Th which has an emission probability of 57.2%. The interference of the 186.2 keV gamma ray resulting from the ^{226}Ra alpha decay to ^{222}Rn, both pertaining to the ^{238}U series, in the determination of ^{235}U is a known fact in gamma spectrometry of soils with naturally occurring radionuclides. Thus, a correction for the ^{226}Ra content of the sample is usually made by taking into consideration the more intense gamma rays from the ^{226}Ra daughters, such as 295.2 keV (18.2%) and 351.9 keV (35.1%) from ^{214}Pb; and 609.3 keV (44.6%), 1,120.3 keV (14.7%), and 1,764.5 keV (15.1%) from ^{214}Bi. This type of correction works properly when radon gas (^{222}Rn) does not escape from the soil sample while being measured. Radon gas leakage is the main shortcoming of such corrective procedures.

Gamma-spectrometric measurements have been improving since the early 1970s. However, the methodology to measure the $^{235}U/^{238}U$ isotopic ratio by gamma spectrometry remained very similar to that described above (Coles et al., 1974) and is still in use today (Bikit et al., 2001; UNEP, 2002; Uyttenhove et al., 2002; IAEA, 2003a, 2003b; Bikit et al., 2003; Paschoa et al., 2003; Paschoa, 2004; McLaughlin, 2005; Bikit et al., 2005; Uyttenhove, 2005) irrespective of the improvements in gamma detection and analytical systems.

Taking into account that it is likely that DU ammunition will continue to be used in wars, modern and more sophisticated methods to measure DU are welcome. In addition, it is desirable that international intercalibration exercises to compare results in standard samples and by different methodologies will become necessary.

As an illustration, Figure 2.1 shows a graph of the linear fit calibration of the $^{235}U/^{238}U$ ratio based on the counting areas under the photopeaks at 143.76 and 63.29 keV, respectively, as a function of the $^{235}U/^{238}U$ standards for DU, NU, and LEW. There are clear advantages of directly measuring the 143.76 keV ^{235}U photopeak (10.96%) instead of the 185.73 keV peak, which is too close of the 186.2 keV gamma ray resulting from the ^{226}Ra alpha decay to ^{222}Rn.

2.2. DU Inventories

The possibility of future military uses of DU is real. One of the main reasons to enhance the probability that DU will continue to be used by the military is the fact that such use is not specifically subject to the Certain Conventional Weapons Agreement (UNOG, 2001) or any other Geneva-based weapons agreements since the 1925 Gas protocol. The reader interested in further discussion of this subject is referred to the

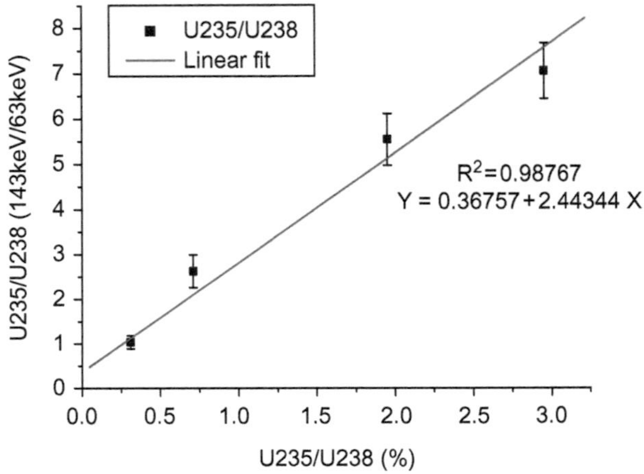

Figure 2.1 Linear fit calibration graph of the $^{235}U/^{238}U$ ratio based on the counting areas under the photopeaks at 143.76 and 63.29 keV, respectively, as a function of the $^{235}U/^{238}U$ standards for DU, NU, and LEW.

intervention made by Avril McDonald at the expert meeting on the Risks of Depleted Uranium Use in Weapon Systems (McDonald, 2008).

Thus, the knowledge of inventories of DU in the world may become important either to negotiate the prevention of using DU in the future or to learn from where the threats of using it may come. Figures 2.2 and 2.3 illustrate how DU wastes are produced in the enrichment (assuming 3.6% ^{235}U for a pressurized water reactor, PWR) and re-enrichment (assuming 3.6% ^{235}U plus 0.3% ^{235}U tails to feed the re-enrichment stream line) processes, respectively (Diehl, 1999).

The large amounts of uranium hexafluoride depleted in ^{235}U (DUF_6) are usually stored in steel cylinders. Although UF_6 is a gas at ambient temperature, it sublimates at 56.4 °C reacting with water (because of humidity) and forms hydrofluoric acid and uranyl fluoride (UO_2F_2). Because of corrosion and toxicity, the steel cylinders may become hazardous. As a consequence, long-term storage of DUF_6 is not advisable. Usually DUF_6 is converted to DUF_4, which can be reconverted to DUF_6, DU_3O_8 (triuranium octaoxide depleted in ^{235}U) which is more stable (because it mimics one of the natural forms of uranium), or to DUO_2 (uranium dioxide depleted in ^{235}U), which will eventually be converted to DU_3O_8 at ambient temperature.

Table 2.2 lists the main physical characteristics of UF_6, UF_4, UO_2F_2, U_3O_8, UO_2, and metallic uranium. DU inventories are maintained in several chemical forms, thus with different physical characteristics.

There are several long-term storage sites of DU throughout the world. This fact may create an environmental impact in the years to come. Table 2.3 lists the world inventory of DU.

FEED 11.9 t UF_6 (8.05 t U natural) 0.71% ^{235}U → ENRICHMENT 4,531 SWU* → PRODUCT 1.48 t eUF_6 (1 t eU) 3.6% ^{235}U

↓

WASTE 10.42 t DUF_6 (7.05 t DU) 0.3% ^{235}U

*Enrichment cost ($ per Separative Work Unit – SWU).

Figure 2.2 Stream of enriched (eUF_6) and depleted uranium (DUF_6) production stream from natural uranium (UF_6). Adapted from Diehl (1999).

FEED 11.9 t UF_6 (8.05 t U natural) 0.71% ^{235}U → ENRICHMENT 4,531 SWU* → PRODUCT 1.48 t eUF_6 (1 t eU) 3.6% ^{235}U

↓

WASTE/REFEED 10.42 t DUF_6 (7.05 t DU) 0.3% ^{235}U → ENRICHMENT 489 SWU → PRODUCT 1.13 t eUF_6 (0.77 t U *natural*) 3.6% ^{235}U

↓

WASTE 9.29 t DUF_6 (6.28 t DU) 0.25% ^{235}U

*Enrichment cost ($ per Separative Work Unit – SWU).

Figure 2.3 Stream of re-enriched (eUF_6) and depleted uranium (DUF_6) production streams from natural uranium (UF_6). Adapted from Diehl (1999).

Table 2.2 Melting point (°C), density ($g\,cm^{-3}$), and solubility in water at ambient temperature of UF_6, UF_4, UO_2F_2, U_3O_8, UO_2, and metallic uranium.

Compound	Melting point (°C)	Density ($g\,cm^{-3}$)	Solubility
UF_6	64.1	4.6	→ UO_2F_2
UF_4	960 ± 5	2.0–4.5	Slightly soluble
UO_2F_2	→ U_3O_8 at 300	~2.6	Soluble
U_3O_8	→ UO_2 at 1,300	1.5–4.0	Insoluble
UO_2	2,878 ± 20	2.0–5.0	Insoluble
Metallic uranium	1,132	19	Insoluble

Source: Adapted from US DoE (2001).
Note: → Means decomposes to

Table 2.3 Known global inventory of depleted uranium with distribution by country, metric tons, chemical form, and percentage.

		DU stocks as reported (t U)					
		UF_6	U_3O_8	Metallic	Total		
USA	DoE plus USEC	4.7×10^5	1.0×10^4		4.8×10^5	2002	40.4
Russia[a]		4.5×10^5	1.0×10^4		4.6×10^5	1996	38.7
France	COGEMA plus EURODIF	5.0×10^4	1.4×10^5		1.9×10^5	2001	16.0
UK	BNFL	3.0×10^4			3.0×10^4	2001	2.5
Germany, the Netherlands plus UK	URENCO	1.6×10^4			1.6×10^4	1999	1.3
Japan	JNFL plus JNC	1.0×10^4			1.0×10^4	2001	0.8
China	CNNC	2.0×10^3			2.0×10^3	2000	0.2
South Korea	KAERI	2.0×10^2			2.0×10^2	2002	0.02
South Africa	NECSA	4	69		73	2001	0.006
Total		1,028,204	160,069		1,188,273		99.926

USEC, former United States Enrichment Corporation; COGEMA, "Compagnie Générale des Matières Nucléaires" which is today AREVA NC; EURODIF, European Gaseous Diffusion Uranium Enrichment Consortium; BNFL, British Nuclear Fuels; URENCO, the joint industrial enterprise established under the Treaty of Almelo; JNFL, Japan Nuclear Fuel Limited; JNC, Japan Nuclear Fuel Cycle Development Institute; CNNC, China National Nuclear Corporation; KAERI, Korea Atomic Energy Research Institute; NECSA, Nuclear Energy Corporation of South Africa.

[a]Estimated in 1996 by the Uranium Institute (today World Nuclear Association, WNA).

As one can see in Table 2.3, UF_6 is (or was until the reported year) the most common chemical form of DU storage. The United States and Russia are responsible for nearly 80% of the total stock of DU reported in the world. France has most of its DU as U_3O_8, which is the most stable form for storage.

2.3. DU as TENR and its Impacts

2.3.1. Main uses of DU

It is well known that there are insoluble uranium oxide particles in uranium mines and in facilities that process nuclear fuels. However, the pyrophoric behavior of metallic uranium particles may produce small hot dust particles which if inhaled may incur internal localized alpha radiation doses in addition to the uranium chemical toxicity.

When DU is used as shielding in casks for a number of materials, including spent fuel or high-level waste, there is scant possibility of forming hot dust particles. However, DU is used (sometimes mixed with other dense materials under the name staballoy) to pierce armor plating, in cruise missiles, and in the armor of tanks, which may lead to the pyrophoric behavior of metallic uranium in the event of battle. In such cases, DU may become a toxic as well as a radiological problem. Some reports indicate that in DU used for military purposes, there are trace contaminants at the parts per billion levels of neptunium, plutonium, americium, technetium-99, and uranium-236. However, at these levels, the contaminants are not of radiological concern.

DU was used as ballast in the tails of the first Boeing 747 aeroplanes. The radiological exposure of bystanders was discussed in the event of an aircraft accident, as occurred in Amsterdam in 1992 (Betti, 2003). It is well known that the 22.5 m ocean racer Pen Duick VI (1973), using a DU ballast keel, won the OSTAR (The Original Single-Handed Trans-Atlantic Race) race of 1976.

2.3.2. WHO's views on DU

In 2001, a WHO document concluded the following (WHO, 2001):

1. a significant impact of DU can be achieved only by military use;
2. levels in contaminated food and ground water could rise after some years of initial DU contamination and should be monitored and appropriate measures taken where there is potential for significant quantities of DU entering the food chain;
3. the WHO guidelines for drinking water quality, 2 μg of uranium per liter, should apply to DU;
4. areas with high concentrations of DU should be cordoned off until cleaned up;

5. disposal of DU should conform to the appropriate recommendations for use of radioactive materials;
6. for soluble DU compounds, limits should be based on a tolerable intake value of 0.5 μg per kg of body weight per day while for insoluble DU compounds, limits should be based on both chemical effects and radiation dose limits as prescribed in the International Basic Safety Standards (BSS) on radiation protection (IAEA, 1996a);
7. general screening or monitoring for possible DU-related health effects is not necessary for populations living in conflict areas where DU has been used;
8. those individuals who believe that they have been exposed to excessive amounts of DU should consult their medical practitioner for treatment of any symptoms and follow-up;
9. young children playing within a former conflict zone could receive a DU exposure due to ingestion of contaminated soil because of hand-to-mouth activity;
10. there are gaps in knowledge – thus further research is recommended in some areas to improve health risk assessment;
11. studies are needed to better understand the extent, reversibility, and possible existence of thresholds for kidney damage in people exposed to DU; and
12. review of studies of populations exposed to naturally elevated levels of uranium in drinking water could yield important information on similar exposures to DU.

2.3.3. IOM's recommendations on DU

Civilian applications of DU do not affect to any appreciable extent the size of the existing stockpiles of DU. As a consequence, the tendency to use DU in military weaponry and shielding is much more significant than the uses of DU in civilian applications. A book published in 2008 by the Institute of Medicine (IOM) of the National Academies updates the epidemiological studies of American veterans exposed to DU (IOM, 2008).

2.3.3.1. Recommendations for epidemiological studies on DU

Concerning the desirable improvement of epidemiological studies on DU, the IOM recommends the following (IOM, 2008):

- DOD should use the most sensitive assays when collecting biomarker data and should investigate available *in vivo* assay techniques other than measuring urinary uranium to determine whether they offer advantages (for example, increased sensitivity) over urinalysis.
- DOD should consider assessing uranium concentrations in lung, kidney, and brain tissues from military personnel who were potentially exposed to DU and died while on active

duty. It is the committee's understanding that DOD maintains a repository of autopsy material from all military personnel who die while on active duty (M. Kilpatrick, personal communication, March 22, 2007, and February 11, 2008). Analysis of uranium in autopsy tissue might provide information on concordance between renal uranium concentrations and model-based estimates, such as those presented in a Royal Society report on the health hazards associated with DU munitions (Royal Society, 2001). It would also provide information on pulmonary retention of DU, which has implications for estimating lung cancer risk, and insight into the toxicokinetics of DU.

- DOD should continue to link and integrate available databases so that information can be assessed. With additional data-linkage capabilities, DOD and VA will be better able to use available data to assess DU exposure and related health outcomes.U.S. Department of Defense (DOD)

3.3.3.2. Further IOM recommendations on DU

There are additional IOM recommendations (IOM, 2008) as follows:

> The committee commends DOD's Depleted Uranium Medical Management Program and recommends that DOD expand its efforts to identify all active-duty military personnel and veterans with potential DU exposure to determine their DU body burdens. Specifically, the committee recommends that:

- DOD determine the feasibility of collecting biomarker data through the Millennium Cohort Study from the 2,823 members of Panel 1 (or potentially from members of all three panels if such information is available) who reported being exposed to DU in the preceding 3 years. That information could be used to conduct an exposure study to assess the accuracy of surrogate exposure measurements (material on questionnaires). Any study participants who test positive for DU exposure should receive health monitoring through the Depleted Uranium Follow-Up Program throughout their lifetime. Comprehensive clinical assessments of exposed people provide extensive data on health outcomes potentially related to DU exposure and might have value in stimulating further study.
- DOD determines the feasibility of identifying military personnel who were at Camp Doha during the time of the fire in 1991 and of collecting biomarker data on them. This work could provide exposure information on level II exposure and level III exposure during the Gulf War. Any veteran who tests positive for DU exposure should receive health monitoring (see above). The committee's final recommendation is related to studying adverse reproductive and developmental outcomes.
- The committee recommends further study of the potential reproductive and developmental toxicity of DU with animal models [a review of reproductive and developmental toxicity studies of DU is included in *Gulf War and Health: Updated Literature Review of Depleted Uranium* (IOM, 2008)]. Specific studies may address reproductive measures (for example, the number of motile sperm, which is highly predictive of intrauterine-insemination success and, thus, correlated with fertilization rate). Important observations on the effects of DU on developing animals might be derived from *in vivo* animal studies. For example, they could provide information on growth, reproductive capacity, cancer, neurobehavioral function, and transplacental exposure. Animal studies focusing on the health effects of DU on the progeny's health might be useful for corroborating findings reported in epidemiological studies, and outcomes may be correlated with DU exposure (as measured by urinary DU excretion).

CHAPTER 3

Terrestrial, Atmospheric, and Aquatic Natural Radioactivity

3.1. Terrestrial and Atmospheric Natural Radioactivity

There are about 340 nuclides in nature, of which about 70 are radioactive. The latter are called the naturally occurring radionuclides, and are found mostly as isotopes of heavy elements. The radioactivity of the natural radionuclides of the uranium (^{235}U and ^{238}U) and thorium (^{232}Th) series is due essentially to alpha and beta decay modes plus gamma deexcitation of nuclei. The beta and electron capture decay modes of ^{40}K to ^{40}Ca (89%) and ^{40}Ar (11%), respectively, the latter followed by the emission of a 1.46 MeV gamma ray, contribute significantly to the natural radioactivity, There are also radiation processes of atomic deexcitation such as X-ray and Auger electron emissions that need to be accounted for as far as natural radiation is concerned.

The mechanisms that determined uranium and thorium distributions through geological times in the Earth's asthenosphere (U, Th richer sphere), mantle, core-mantle boundary and outer core have been discussed by Bao (2008). The concentrations of ^{40}K, ^{232}Th, and ^{238}U in the Earth's crust are assumed to have increased throughout geological times (Karam and Leslie, 2005). However, taking into account that during the last 2 Ga these three radionuclides have kept decaying, the annual external radiation dose from the terrestrial nuclides present in the Earth's crust is believed to have been maintained constant at about 1.6 mGy y^{-1} during this same period.

The nuclides of the ^{238}U, ^{235}U, and ^{232}Th radioactive decay series are shown in Tables 3.1–3.3 with their historical names, half-lives, and main radiation.

The ranges and average concentrations of ^{40}K, ^{232}Th, and ^{238}U in typical rocks and soils in the Earth's crust are shown in Table 3.4. Much of the external natural background gamma radiation comes from the two main radioactive series (^{232}Th and ^{238}U) and the 1.46 MeV gamma from ^{40}K.

However, the contribution of muons to the external radiation measured with the common radiation detectors is usually misinterpreted as terrestrial gamma radiation. The typical radiation detector converts ionization produced by ionizing radiation into some kind of detectable signal. In the

Radioactivity in the Environment, Volume 17
ISSN 1569-4860, DOI 10.1016/S1569-4860(09)01703-3

Table 3.1 Decay series of ^{238}U[a].

Nuclide	Historical name	Half-life	Main radiation
^{238}U	Uranium I	4.47×10^9 y	α, <1%[b] γ
^{234}Th	Uranium X_1	24.1 d	β
^{234m}Pa	Uranium X_2	1.17 m	β, <1% γ
^{234}U	Uranium II	2.46×10^5 y	α, <1% γ
^{230}Th	Ionium	7.54×10^4 y	α, <1% γ
^{226}Ra	Radium	1.6×10^3 y	α, γ
^{222}Rn	Emanation	3.82 d	α, <1% γ
^{218}Po	Radium A	3.10 m	α, <1% γ
^{214}Pb	Radium B	26.8 m	β, γ
^{214}Bi	Radium C	19.9 m	β, γ
^{214}Po	Radium C'	164.3 μs	α, <1% γ
^{210}Pb	Radium D	22.3 y	β, γ
^{210}Bi	Radium E	5.01 d	β
^{210}Po	Radium F	138.4 d	α, <1% γ
^{206}Pb	Radium G	Stable	None

[a]NuDat online by Brookhaven National Laboratory – http://www.nndc.bnl.gov/nudat2/reCenter.jsp?z = 90&n = 142
[b]Minor branches, < 1%, not shown.

Table 3.2 Decay series of ^{232}Th[a].

Nuclide	Historical name	Half-life	Main radiation
^{232}Th	Thorium	1.41×10^{10} y	α, <1%[b] γ
^{228}Ra	Mesothorium I	5.75 y	β, <1% γ
^{228}Ac	Mesothorium II	6.15 h	β, γ
^{228}Th	Radiothorium	1.91 y	α, γ
^{224}Ra	Thorium X	3.66 d	α, γ
^{220}Rn	Emanation (thoron)	55.6 s	α, <1% γ
^{216}Po	Thorium A	0.145 s	α, <1% γ
^{212}Pb	Thorium B	10.64 h	β, γ
^{212}Bi	Thorium C	1.01 h	α, γ
^{212}Po (64%) – ^{208}Tl (36%)	Thorium C′ – Thorium C′	0.3 ms – 3.05 m	α – β, γ
^{208}Pb	Thorium D	Stable	None

[a]NuDat online by Brookhaven National Laboratory – http://www.nndc.bnl.gov/nudat2/reCenter.jsp?z = 90&n = 142.
[b]Minor branches, < 1%, not shown.

Table 3.3 Decay series of ^{235}U (actinium)[a].

Nuclide	Historical name	Half-life	Main radiation
^{235}U	Actiniuranium	7.04×10^8 y	α, γ
^{231}Th	Uranium Y	1.06 d	β, γ
^{231}Pa	Protoactinium	3.28×10^4 y	α, γ
^{227}Ac	Actinium	21.77 y	β, <1%[b] γ
^{237}Th (98.6%) – ^{223}Fr (1.4%)	Radioactinium – Actinium K	18.7 d – 22.0 m	α, γ – β, γ
^{223}Ra	Actinium X	11.44 d	α, γ
^{219}Rn	Actinon	3.96 s	α, γ
^{215}Po	Actinium A	1.78 ms	α, <1% γ
^{211}Pb	Actinium B	36.1 m	β, γ
^{211}Bi	Actinium C	2.14 m	α, γ
^{207}Tl	Actinium C′	4.77 m	β, <1% γ
^{207}Pb	Actinium D	Stable	None

[a]NuDat online by Brookhaven National Laboratory – http://www.nndc.bnl.gov/nudat2/reCenter.jsp?z=90&n=142.
[b]Minor branches, <1%, not shown.

Table 3.4 Ranges and average concentrations of ^{40}K, ^{232}Th, and ^{238}U in typical rocks and soils.

	Bq kg^{-1}		
	^{40}K	^{232}Th	^{238}U
Igneous rocks	3.0×10^2–$>1.0 \times 10^3$	10–70	7–40
Sedimentary rocks	$<3.0 \times 10^2$–9.0×10^2	<8–50	10–40
Beach sands[a]	$<3.0 \times 10^2$	25	40
Carbonate rocks	70	8	25
All rocks	70–1.4×10^3	7–80	7–60
Continental crust	8.5×10^2	44	36
Soil	4.0×10^2	37	22

Source: Adapted from NAS-NRC (1999).
[a]Some beaches, like monazite-bearing beach sands, have typical ^{232}Th plus progeny and ^{238}U plus progeny activity concentrations of the order of 10^4 Bq kg^{-1} and 10^3 Bq kg^{-1}, respectively (Paschoa and Dias da Cunha, 2008).

case of alpha and beta particles, ionization is produced by direct collision with the detector material. For gamma radiation there are three main processes to be considered: the photoelectric effect, Compton scattering, and pair production. Most gamma radiation detectors take advantage of at least one of those three processes to obtain a detectable signal. A scintillation detector for gamma radiation is always associated with a photomultiplier tube.

A proportional counter, for example, is just an ionization chamber with an applied high-voltage bias to yield electrons from ionization processes with energies high enough to produce secondary electrons. The amplifying factor of a proportional counter is of the order of 10^4 or even higher. In general, a proportional counter is used with a stable high-voltage power supply. Thus, a proportional counter can integrate all external radiation, which reaches the effective area (or volume) of the detector.

When one uses a proportional counter or a scintillation detector, gamma radiation is detected, as well as muons. The latter, however, are not always considered, or admitted only as a minor fraction of the external radiation detected. Accordingly, measurements of external natural radiation with proportional counters or NaI(Tl), for example, include muons that are not properly considered in dosimetric estimates. The dose factors used for dose estimates from external natural radiation based on the most common measurement detectors need to be reevaluated.

Muons are produced in interactions that occur in the high atmosphere between nuclei of gaseous molecules and primary cosmic rays, which are mostly high-energy protons. As a result of these interactions, secondary cosmic rays are produced. A large fraction of these secondary cosmic rays comprises pions, which are very short-lived mesons. The three types of pions are the following: π^0 (neutral); π^+; *e* π^-. The latter two are the particle and the antiparticle of each other. While the decay of π^0 produces two protons, which in turn produce other particles, the two charged pions decay as follows:

$\pi^+ \rightarrow \mu^+ + \nu_\mu$ producing an anti-muon and a muon-neutrino and
$\pi^- \rightarrow \mu^- + \nu_\mu$ producing a muon and a muon-anti-neutrino.

There are two important characteristics of the relativistic muons worth mentioning here. First they have half-lives of the order of 2 μs, which is much longer than those of pions. The π^0, for example, has a half-life of 8×10^{-15} s. Second, because muons do not undergo many interactions, they easily reach the Earth's surface. As a consequence, muons represent the vast majority of the cosmic background radiation that reaches sea level.

As a corollary of these characteristics of muons, there is a high percentage of the muon component in the hadronic shower when external radiation is measured offshore, far from soil and rocks (de Freitas et al., 2007). Inland the local geological configuration and the radon concentration near the detector contribute significantly to the gamma component of the external radiation measurements, because of the gamma emissions of the members of the natural series of ^{238}U e ^{232}Th, and the 1.46 MeV gamma of the ^{40}K. In particular, the contribution of the gamma emissions of ^{214}Bi, which is part of the ^{222}Rn progeny, increases the number of events registered by radiation detectors sensitive to gamma radiation. The terrestrial gamma radiation is measured in the range of 0–3 MeV, thus including the 2.614 MeV gamma emission of ^{208}Tl, thus not registering in the gamma spectra most of the gamma energies originating from the cosmic radiation component.

Table 3.5 Dose coefficients for external gamma radiation coming from soil.

	Dose coefficient ($nGy\,h^{-1}$ per $Bq\,kg^{-1}$)		
	UNSCEAR (1977), Beck (1972)	ICRU (1994), Saito and Jacob (1995), Saito et al. (1998), UNSCEAR (2000)	Eckerman and Ryman (1993)
^{40}K	0.043	0.0417	0.046
^{238}U and progeny	0.427	0.462	0.561
^{232}Th and progeny	0.662	0.604	0.682

Table 3.5 shows the evolution of dose coefficients as reported by the United Nations Scientific Committee on the Effects of Atomic Radiation (UNSCEAR) – in $nGy\,h^{-1}$ per $Bq\,kg^{-1}$ – and other publications. The dose coefficients listed in Table 3.5 have been used worldwide throughout the years since published. The slight differences in dose coefficients presented in the three columns of Table 3.5 are not taken into consideration in most calculations. Eckerman and Ryman (1993), however, called attention to the fact that significant uncertainties in the average composition of soil may result in quite different dose coefficients, and as a result different dose estimates.

The conversion from activity concentrations in soil to absorbed dose rates uses slightly different coefficients, as illustrated in Table 3.5. A conversion from activity concentrations in soil to absorbed dose rates is usually made by a simple formula:

$$D(n\mathrm{Gy\,h^{-1}}) = C_{\mathrm{Th}}S_{\mathrm{Th}} + C_{\mathrm{U}}S_{\mathrm{U}} + C_{\mathrm{K}}S_{\mathrm{K}} \tag{3.1}$$

where C_{Th}, C_{U}, and C_{K} are the dose coefficients for Th, U, and K, respectively – taken from one of the columns of Table 3.1.5 – and S_{Th} is the measured activity concentration measured through the ^{208}Tl photo-peak (in $Bq\,kg^{-1}$), S_{U} is the activity concentration measured through the ^{214}Bi photo-peaks (in $Bq\,kg^{-1}$), and S_{K} is the activity concentration measured through the ^{40}K photo-peak (in $Bq\,kg^{-1}$).

On the contrary, the relative contributions to the terrestrial gamma dose rate D from Th, U, and K (R_{Th}, R_{U}, and R_{K}) in soil were estimated from gamma-ray dose rates using the following formula (Minato, 2006; Yamada et al., 2008):

$$R_{\mathrm{Th}} = \frac{X_{\mathrm{Th}}C_{\mathrm{Th}}}{D}, R_{\mathrm{U}}\frac{Y_{\mathrm{U}}C_{\mathrm{U}}}{D}, R_{\mathrm{K}} = \frac{Z_{\mathrm{K}}C_{\mathrm{K}}}{D} \tag{3.2}$$

where $X_{\mathrm{Th}} = 2.7$, $Y_{\mathrm{U}} = 5.4$, $Z_{\mathrm{K}} = 13.0$, and C_{Th}, C_{U}, C_{K} are, respectively, the concentrations of Th, U, and K in soil.

The percentage errors on Th, U, and K estimated using Equation (3.2) were reported to be 8, 11, and 4%, respectively. These relative estimates and corresponding errors are based on the geological formations under the soils measured. The X_{Th}, Y_U, and Z_K values used in Equation (3.3) reflect the geological formation at specific sites in Japan. However, when there are technological enhancements of one or more of the natural radionuclides in soils, the values of X_{Th}, Y_U, and Z_K may be significantly different.

It is worth comparing in Table 3.6 a soil technologically enhanced with natural radionuclides with typical values of ^{40}K, ^{238}U plus progeny, and ^{232}Th plus progeny reported in the open literature. The disequilibria of the ^{238}U and ^{232}Th series generated by industrial processes make it important to report the activity concentration of individual natural radionuclides. Enhancements in activity concentrations of 6-fold for ^{238}U and 10-fold for ^{232}Th contrast with a decrease of more than 20-fold for ^{40}K activity concentration. The use of an equation equivalent to Equation (3.2) requires, in the case of soils contaminated with natural radionuclides, further research.

It is worth remembering that soils from different domains will have been subject to distinct weathering processes during time, these quite likely dependent on geomorphology, vegetation coverage, long- and short-term climatic conditions and age, just to mention a few factors.

Table 3.6 Typical values of activity concentrations of ^{40}K, ^{238}U plus progeny, and ^{232}Th plus progeny, in soils and those values for ^{40}K, ^{238}U, ^{226}Ra, ^{210}Pb, ^{228}Ra, and ^{228}Th reported in red mud[a].

	$Bq\,kg^{-1}$		
	Typical activity concentration in soil		Activity concentration in red mud
	NCRP (1987)	UNSCEAR (2000)	
^{40}K	4.0×10^2	4.0×10^2	< 20
^{238}U plus progeny	22	33	–
^{238}U	–	–	190 ± 40[b]
^{226}Ra	–	–	190 ± 30
^{210}Pb	–	–	150 ± 30
^{232}Th plus progeny	37	30	–
^{228}Ra	–	–	370 ± 50
^{228}Th	–	–	380 ± 50

[a]Red mud is an unwanted by-product of the industrial process to produce aluminum oxide from bauxite – it is listed as enhanced natural radiation material under the RPO (German Radiation Protection Ordinance (RPO) – Verordnung über dan Schutz vor Schäden derch ionisierend Strahlen (Strahlen-schutzverordnung) – Strisch V (2001) Bundesgesetzblatt I, 1714) rules.

[b]2σ uncertainties.

It is important to know that the behavior of each naturally occurring radionuclide varies according to a number of factors, for example (see NAS-NRC, 1999):

380. different specific radionuclide mobilities along the decay chains;
381. changes in the physical and chemical forms of the radionuclides due to technological and industrial processes and released into a variety of environmental media;
382. temporal variation of the naturally occurring radioactive materials (NORM) due to weathering reactions; and
383. changes in activity concentrations and physical and chemical forms of the naturally occurring natural radionuclides.

K-40 is the only radioactive potassium isotope. Its half-life is 1.3×10^9 y. The ^{40}K contents in soil and rocks range from 70 to 1,400 Bq kg^{-1} (see Table 3.4). Some basalts and sands are low in ^{40}K while other basalts and granites have very high content. K-40, however, is not the only non-series natural radionuclide that is present everywhere on Earth. The primordial radionuclide ^{87}Rb is present in small quantities ($\approx$70 Bk kg^{-1}) in the Earth's crust. The half-life of ^{87}Rb is 4.75×10^{10} y. It is known that ^{87}Rb is present in the environment and also in human tissues, delivering an average annual effective dose equivalent between 3 and 6 μSv (NCRP, 1987; UNSCEAR, 1988; NAS-NRC, 1999).

In addition to ^{40}K and ^{87}Rb, there are other non-series natural radionuclides, but all of them have very long half-lives ($>10^{10}$ y). As a consequence, these non-series radionuclides have very low activity concentrations in the Earth's crust, and do not have any importance from the radiological viewpoint. Table 3.7 lists the non-chain primordial radionuclides, including ^{40}K and ^{87}Rb, their half-lives, major radiation emissions, and typical activity concentrations in the Earth's crust. With the exception of ^{40}K and, to a lesser extension, ^{87}Rb, whose typical Earth's crustal activity concentrations are 630 and 70 Bq kg^{-1}, respectively, all other non-chain radionuclides have no importance whatsoever from the radiological viewpoint because of their characteristics.

Both ^{40}K and ^{87}Rb, however, have been used in geochronology studies since the 1930s. In 1937, von Weizsäcker observed that the inert gas argon was much more abundant than helium, neon, krypton, and xenon together, and suggested that ^{40}Ar was the result of the electron capture decay of ^{40}K (von Weizsäcker, 1937). After WWII, the K–Ar method for age determination of terrestrial materials became popular (see e.g., Smits and Gentner, 1950; Wasserburg and Hayden, 1955; Wetherhill et al., 1955). New techniques today allow the use of the K–Ar method in geochronology in association with other methods (see e.g., Oliveros et al., 2008).

Considering that the rubidium in mica (leptodolites) and feldspar is a few percent, and that the strontium isobar accumulates over geological time

Table 3.7 Naturally occurring non-chain primordial radionuclides with half-lives, major radiation emissions, and typical Earth's crustal activity concentrations[a].

Radionuclide	Half-life (y)	Major radiation	Typical Earth's crustal activity concentration ($Bq\,kg^{-1}$)
^{40}K	1.28×10^{9}	β, γ	630
^{87}Rb	4.75×10^{10}	γ	70
^{147}Sm	1.06×10^{11}	α	0.7
^{176}Lu	3.73×10^{10}	β, γ	0.04
^{138}La	1.05×10^{11}	β, γ	0.02
^{187}Re	4.3×10^{10}	β	1×10^{-3}
^{144}Nd	2.29×10^{15}	α	3×10^{-4}
^{50}V	1.4×10^{17}	γ	2×10^{-5}
^{115}In	6×10^{14}	β	2×10^{-5}
^{142}Ce	$>5 \times 10^{16}$	β	$<1 \times 10^{-5}$
^{152}Gd	1.08×10^{14}	α	7×10^{-6}
^{113}Cd	9×10^{15}	β	$<2 \times 10^{-6}$
^{123}Te	1.24×10^{13}	X-rays	2×10^{-7}
^{174}Hf	2.0×10^{15}	α	2×10^{-7}
^{190}Pt	6.5×10^{11}	α	7×10^{-8}

[a]http://www.nndc.bnl.gov/nudat2/reCenter.jsp?z = 90&n = 142

by the decay

$$^{87}Rb \rightarrow {}^{87}Sr + \beta \tag{3.3}$$

the possibility of determining geological ages of mineral such as mica and feldspar was suggested in the late 1930s (Hahn and Walling, 1938; Ahrens, 1940). Today the Rb–Sr technique is quite popular in geochronology (see e.g., Schofield et al., 2008).

3.1.1. Uranium

The primordial uranium isotope ^{238}U is found ubiquitously in nature, as indicated by the data in Table 3.4. However, as described in detail in the previous chapter, ^{238}U (4.47×10^{9} y half-life) constitutes 99.27% by mass of the uranium found in the Earth's crust, the remaining naturally occurring uranium isotopes being ^{235}U (7.04×10^{8} y half-life), which is the parent isotope of the actinium chain (see Table 3.3), and ^{234}U (2.46×10^{5} y half-life), which is the fourth member of the ^{238}U chain (see Table 3.1).

Activity concentrations reported for ^{238}U in soils have a wide range from 2 to 690 $Bq\,kg^{-1}$ (UNSCEAR, 2000). On the contrary, the activity concentrations of ^{238}U in rocks have a much narrower range from 7 to

$60\,Bq\,kg^{-1}$ (see Table 3.4). There are, however, veins in some specific rocks that contain uranium at over 100 ppm ($1.2 \times 10^3\,Bq\,kg^{-1}$). A phosphorus ore that contains uranium at 120 ppm (1.4×10^3) is a commercial source of uranium (NCRP, 1993). Uranium ores are those natural materials which contain over 500 ppm ($6.0 \times 10^3\,Bq\,^{238}U\,kg^{-1}$) (NAS-NRC, 1999).

3.1.2. Radium-226

The sixth member of the ^{238}U decay chain (see Table 3.1) is the radium isotope ^{226}Ra (1.6×10^3 y half-life). This radionuclide alpha-decays into ^{222}Rn (3.82 d half-life), which is a noble gas. Beginning with ^{222}Rn, a series of six short-lived alpha and beta disintegrations occurs (see Table 3.1) leading to the stopper ^{210}Pb (22.3 y half-life). This latter nuclide decays by beta emission to ^{210}Bi (5.01 d half-life), which in turn decays by beta particle into ^{210}Po (138.4 d half-life). The last member of the ^{238}U decay chain is the stable ^{206}Pb, which is the result of the alpha decay of ^{210}Po. There are several gamma-emitting nuclides in the ^{226}Ra progeny (see Table 3.1).

Activity concentrations reported for ^{226}Ra in soils have a wide range from 1 to $900\,Bq\,kg^{-1}$ (UNSCEAR, 2000). However, one cannot be sure whether some high ^{226}Ra content soils were earlier contaminated with TENORM, or not. There are numerous historical orphan sources throughout the world, which may become mixed with previously uncontaminated soils. The fact that many uranium ores have activity concentrations much higher than those reported in UNSCEAR for ^{226}Ra suggests that there must be some soils contaminated with orphan sources of TENORM with activity concentrations of ^{226}Ra of the order of $10^3\,Bq\,kg^{-1}$.

3.1.3. Radon

The mechanism through which the ^{222}Rn isotope of the noble gas radon is transported from soil to the atmosphere, as well as radon measurement techniques, have been extensively studied for a long time now (see e.g., Lucas, 1964; Lockart, 1964; Tanner, 1964, 1980; Tanner and Olsen, 1992). Activity concentrations of ^{222}Rn outdoors vary from <1 up to a few tens of $Bq\,m^{-3}$ (Gesell, 1983; NAS-NRC, 1999). The subjacent geological formation is a determining factor for outdoor activity concentrations of ^{222}Rn. Outdoor radon activity concentrations, however, vary diurnally as a function of the time of day, temperature, atmospheric pressure, humidity, and with exhalation rate from soil (see e.g., Äkerblom, 1986; Wilkening, 1990; Mahesh et al., 2005; Kozak et al., 2005). There might be some places with outdoors radon higher than $100\,Bq\,m^{-3}$, as was the case for activity concentrations near the uranium mill tailings near Salt Lake City, UT in winter periods when temperature inversion occurred, trapping radon gas emitted from the tailings (Paschoa, 1984, personal communication; Paschoa et al., 1985), but such high outdoors ^{222}Rn is likely to be due to anthropogenically accumulated ^{226}Ra. TENORM outdoor radon is usually associated with TENORM ^{226}Ra.

In the 1970s and 1980s, the view that natural radiation did not have significant radiological consequences changed when it was discovered that a large number of houses in several countries had indoor radon levels of the order of $10^3\,Bq\,m^{-3}$ and even higher (Steinhäusler, 1975; George and Breslin, 1980; Gunning and Scott, 1982; Gesell, 1983; Castrén et al., 1985; McLaughlin, 1988; Swedjemark et al., 1993). The more interested reader on the historical accounts of radon research measurements and hazards can consult a number of interesting articles (see e.g., Miller, 1990; Schmalz, 1990; Swedjemark, 2004; Lewis, 2006). There is also a recent review of the chronological history of radon research and measurements (George, 2008).

In accordance with UNSCEAR (2000), indoor ^{222}Rn activity concentrations range from $24\,Bq\,m^{-3}$ in Egypt (Kenaway and Morsy, 1991) to about $8.5 \times 10^4\,Bq\,m^{-3}$ (Swedjemark, et al., 1993). The reasons for this wide range of indoor ^{222}Rn activity concentrations are numerous. As in the case of outdoor radon, the subjacent geological formation is a major contributing factor. Construction materials with high and enhanced ^{226}Ra concentrations are also an important contributor to high ^{222}Rn activity concentration indoor. There are many publications on indoor ^{222}Rn in the open literature. In the mid-1980s, overviews were made on the distribution of airborne ^{222}Rn concentrations in US and Swedish houses (Nero et al., 1986; Nero, 1988; Samuelson, 1988).

Table 3.8 summarizes some of the most important contributions of various sources to indoor ^{222}Rn.

Table 3.8 Contributions from various sources to indoor ^{222}Rn activity concentrations – in single-family houses, and in high-rise apartments.

	$Bq\,m^{-3}$	
	Single-family houses	Apartments (high-rise)
Soil[a] (based on flux measurements)	55	>0
Water (from public supplies)	0.4[b]	0.37[b]
Building materials	2	3.7
Outdoor air	10	9.25
Observed indoor concentrations	55	12?

Source: Adapted from Nero et al. (1986), Nero (1988), and NAS-NRC (1999).
[a]Subjacent geological formation may contribute as a soil source.
[b]Contributions may be much larger in some cases due to several factors.

Soil and walls are the largest contributors to indoor ^{222}Rn activity concentration. Radon as a gas exhales from the interstitial spaces of soils and walls to enter a room. Numerous models and procedures have been developed to decrease indoor ^{222}Rn, but the legal limits vary from country to country, and from Europe to the United States (Jacobi, 1972; Wicke and Postendörfer, 1982; Paschoa et al., 1984; Rogers and Nielson, 1991; van der Pal et al., 2005).

The WHO a couple of decades ago started a radon investigation initiative. This led to the Indoor Quality Report (WHO, 1996), which was one of the results of the International Radon Project (IRP). The IRP analyzed radon case control studies from Europe, North America, and China. The WHO report on indoor quality recommended, among other things, that, "Radon policy should be investigated towards a global risk management strategy, for which the various policy tools are developed in a consistent way" (WHO, 1996). One of the conclusions of the WHO investigations is that radon gas is the second most important cause of lung cancer – smoking being the first one (see e.g., WHO, 2004).

There are countries, like Italy, which have established long-term studies on radon exposure of the population and which use these as the basis for national radon policy (Bochicchio et al., 1992, 1994). The US Environmental Protection Agency (EPA) maintains an interactive radon map in a Web site to assist national, state, and local organizations to target their resources and to implement radon-resistant building codes (see http://www.epa.gov/radon/zonemap.html). In the last few years, Europe has started a discussion on the need for, and methods of preparation of, a radon map for Europe (see (i) the Proceedings of the 8th International Workshop on the Geological Aspects of Radon Risk Mapping, held in Prague, Czech Republic, 26–30 September 2006 – http://www.radon-vos.cz; (ii) the Proceedings of the 9th International Workshop on the Geological Aspects of Radon Risk Mapping, held on August 12 and 13, 2008 at the 33rd International Geological Congress, in Oslo, Norway, August 6–14, 2008 – to be published in a special 2009 issue of the *Journal of Environmental Radioactivity* – http://radonmapping.jrc.it). The final European indoor radon grid is expected to be produced on or soon after the 10th International Workshop on the Geological Aspects of Radon Risk Mapping to be held in Prague, Czech Republic in the Fall of 2009.

The ^{222}Rn concentration in soil C_{Rn} in the absence of radon transport is given by the following equation (Nazaroff et al., 1988; Washington and Rose, 1992; UNSCEAR, 2000):

$$C_{Rn} = C_{Ra} f \rho_s e^{-1}(1 - \varepsilon)(m[K_T - 1] + 1)_{-1} \tag{3.4}$$

where C_{Ra} is the activity concentration of ^{226}Ra in soil (Bq kg^{-1}), f the emanation factor, ρ_s the density of the soil grains (typically 2.7×10^3 kg m^{-3}),

ε the total porosity (including water and air phases), m the fraction of the porosity that is water-filled (also called the fraction of saturation), and K_T is the partition coefficient for radon between the water and air phases.

In the case of dry soil, $m=0$, and the last term of Equation (3.4) inside parenthesis can be omitted. Washington and Rose (1992) made a test application of Equation (3.4) for a warm, moist soil (25°C, $K_T=0.23$, $m=0.95$) with soil parameters ($C_{Ra}=30\,Bq\,kg^{-1}$, $f=0.2$, $\varepsilon=0.25$), and found $C_{Rn}=78\,kBq\,m^{-3}$. This result was 3.7 times higher than for the same soil under cold and dry conditions (i.e., 0°C, $K_T=0.53$, $m=0.05$), which means $C_{Rn}=21\,kBq\,m^{-3}$.

The flux density of radon J_D ($Bq\,m^{-2}\,s^{-1}$) at the surface of dry soil for a porous mass of homogeneous material semi-infinite in extent is given by the following equation (UNSCEAR, 1988, 1993):

$$J_D = C_{Ra}\lambda_{Rn} f \rho_s (1-\varepsilon) L \tag{3.5}$$

where C_{Ra} is the activity concentration of ^{226}Ra in the earth ($Bq\,kg^{-1}$), λ_{Rn} the decay constant of ^{222}Rn ($=2.1\times10^{-6}$ s), f the emanation factor for each material, ρ_s the density of the soil grains (typically 2.7×10^3 kg m^{-3}), ε the total porosity of dry earth material, and L the diffusion length $(D_e/\lambda_{Rn})^{1/2}$, with D_e ($m^2\,s^{-1}$) being the diffusion coefficient in the earth material.

Here it is worth warning that Equation (3.5) is valid only for dry soil. Using characteristic parameters (i.e., $C_{Ra}=40\,Bq\,kg^{-1}$, $f=0.2$, $\varepsilon=0.25$, $D_e=2\times10^{-6}\,m^2\,s^{-1}$) in Equation (3.5), one determines $J_D=3.3\times10^{-2}$ $Bq\,m^{-2}\,s^{-1}$ (UNSCEAR, 2000).

Wilkening et al. (1972) have in the past estimated a mean worldwide ^{222}Rn flux $J_D=1.6\times10^{-2}\,Bq\,m^{-2}\,s^{-1}$. Recently (Goto et al., 2008), average global exhalation rates were found to be: (i) $J_D=1.75\times10^{-2}$ Bq $m^{-2}\,s^{-1}$ when the soil water saturation was calculated from precipitation and atmospheric temperature, and (ii) $J_D=1.68\times10^{-2}\,Bq\,m^{-2}\,s^{-1}$ based on the meteorological dataset from the Japan Meteorological Agency. The distribution of the exhalation rate for ^{222}Rn as a function of latitude shows clear peaks near parallels 20°N and 20°S (Schery and Wasiolek, 1998; Goto et al., 2008). It is interesting and worth mentioning here that back in 1962 averages of radon fluxes measured in Lincoln, MA and on the Rainier Mesa at the Nevada Test Site ranged from $J_D=1.5\times10^{-2}\,Bq\,m^{-2}\,s^{-1}$ up to $2.6\times10^{-2}\,Bq\,m^{-2}\,s^{-1}$, depending on wind intensity and barometric conditions (Kraner et al., 1964).

Radon fluxes around $J_D=1.7\times10^{-2}\,Bq\,m^{-2}\,s^{-1}$ with small variations as a function of latitudinal distribution are in good agreement with measurements and estimates made by several investigators (Kraner et al., 1964; Wilkening et al, 1972; Schery and Wasiolek, 1998; Conen and Robertson, 2002; Sun et al., 2004; Goto et al., 2008). It becomes clear that the UNSCEAR overestimates J_D (i.e., $J_D = 3.3 \times 10^{-2}$ Bq m^{-2} s^{-1}). Thus this value needs to be revised either by changing the characteristic parameters or by averaging the estimates made throughout the world.

Back in 1995, the natural background exposure to radon was not included in the dose limit for members of the public (Clarke, 1995). High levels of radon in homes were treated as an intervention situation in accordance with ICRP recommendations (ICRP, 1993). Historically, the general principle of exemption of NORM sources, including radon, and practices from regulatory control adopted by the International Atomic Energy Agency (IAEA) in 1988 were based on the following (IAEA, 1988; Linsley, 1999):

1. to establish exemption levels based on other than trivial dose levels – but the rationale for this was not apparent; and
2. to allow exemption based on optimization considerations in which compliance with any relevant dose constraints, and/or regulatory requirements would not lead to any improvement in radiological protection.

Differently from the application of the concept of exemption for NORM adopted by the IAEA, the EU Directive (EC, 1997) does not apply to exposure to radon in dwellings or to natural levels of radiation – that is to radionuclides contained in the human body, to cosmic radiation prevailing at ground level or above-ground exposure to radionuclides present in the Earth's crust (EC, 1997; Janssens, 1999). The EU prefers to use the concept of exclusion only to exposures which are unamenable to control.

A high activity concentration of indoor ^{222}Rn is usually considered as TENORM. The global effort initiated by WHO in 1996 (WHO, 1996) and current initiatives to construct geological radon risk maps in several parts of the world will tend to gain momentum in the next decade.

Investigations of nontechnological enhancement of radon (NTERn) underground have started many years ago as tentative efforts to predict volcanic eruptions and earthquakes (see among others – Ulomov and Mavashev, 1971; Chirkov, 1975; Smith et al., 1976; King, 1981; Wollenberg et al., 1984; Bella and Pettinelli, 1989; Segovia et al., 1989; Ramola et al., 1990; Khan et al., 1990; Igarashi et al., 1995). At the beginning of the 21st century, attempts to predict earthquakes based on neural networks associated with radon concentration and other environmental parameters (Negarestani et al., 2001, 2003) model the relationships between earthquake magnitude, epicentral distance, and precursor time based on radon anomalies (Planinić et al., 2004), and actual measurements of radon contents of underground water and other parameters like pH, electrical conductivity, pressure of dissolved gases and groundwater temperature (Plastino et al., 2002). More recently, a number of reports have been published in which there was evidence of radon anomalies linked to actual earthquakes (Crokett et al., 2005; Walia et al., 2005, 2006; Das et al., 2005; Gosh et al., 2007).

There was a quite recent claim made by an Italian seismologist, Giampaolo Giuliani, who predicted an earthquake – probably on March 31, 2009 – which actually hit hard the Abruzzo region in central Italy around the town of L'Aquila – on April 6, 2009 – based on the radon anomaly detected at the National Physical Laboratory of Gran Sasso (*Corriere Della Sera*, April 6, 2009; *The Guardian*, April 6, 2009; *The New York Times*, April 6, 2009; see also Hooper, 2009). Apparently, it will take some time until it will be possible to assign with high precision the date and time of an earthquake.

3.1.4. Thorium

As was the case for ^{238}U, ^{232}Th (1.41×10^{10} y half-life) is also a primordial nuclide found ubiquitously in nature, as can be noted from the data listed in Table 3.4. There are shorter-lived thorium isotopes in all three natural decay chains, as follows: ^{234}Th (24.1 d half-life) and ^{230}Th (7.54×10^4 y half-life) in the ^{238}U chain; ^{228}Th (1.9 y half-life) in the ^{232}Th chain; and ^{231}Th (1.06 d half-life) in the ^{235}U chain. There are also a number of gamma-emitting nuclides in the ^{232}Th decay chain. Activity concentrations reported for ^{232}Th in soils also have a wide range from 1 to 220 Bq kg^{-1} (UNSCEAR, 2000). Similarly as for ^{238}U, the activity concentrations of ^{232}Th in rocks, however, have a narrower range from 7 to 80 Bq kg^{-1} (see Table 3.4). Some regions of the world (i.e., the areas of high natural radioactivity) have much higher ^{232}Th plus progeny activity concentrations.

According to Table 3.9, the sites with highest ^{232}Th activity concentrations are located on the Chhatrapur beach, Orissa, India (1.2×10^6–3.2×10^6 Bq kg^{-1}), and Morro do Ferro, Poços de Caldas, Minas Gerais, Brazil (2.3×10^3–4.1×10^4 Bq kg^{-1}). These two sites were quite likely disturbed by anthropogenic activities before being measured for ^{232}Th activity concentrations. The relation between areas of high natural radioactivity and TENORM are today well established (Paschoa and Nóbrega, 1981; Paschoa, 2000a, 2000b; Paschoa and Godoy, 2002).

One can note that part of the Kerala State and the adjoining area in Madras State, both in India, were considered areas of high natural radiation background but their monazite-bearing sand beaches were used for extraction and industrial processing of rare earth concentrates plus other elements. As part of the old local procedures, the finely divided by-product sands were loaded on barges and dumped into the sea a few miles offshore. As time passed, the currents and surf returned the finely divided sand to nearby beaches. One of the interesting consequences of this long-term anthropogenic activity was the fact that ^{226}Ra and ^{228}Ra leaching experiments indicate that monazite sands from Kerala and Madras leach

Table 3.9 List of selected locations in the world with high and very high ^{232}Th activity concentrations.

Location	^{232}Th concentration ($Bq\,kg^{-1}$)	Reference(s)
Chhatrapur beach[a], Orissa, India	1.2×10^6–3.2×10^6	Sengupta et al. (2005)
Morro do Ferro, Poços de Caldas, Minas Gerais, Brazil	2.3×10^3–4.1×10^4	Roser et al. (1964); Paschoa (2000)
Yangjiang, Guangdong, China	10–1.8×10^3	Wang (2002)
Ramsar, Iran	16–120	Sohabi and Esmaili, 2002
Goa, India	2.2–87	Avadhami et al. (2005)

[a]It is quite likely that the extreme high values of ^{232}Th activity concentrations are the result of technological enhancement (see e.g., Paschoa et al., 1993).

these radium isotopes much more than the Brazilian monazite beach sands, which had not undergone equivalent anthropogenic activities (Paschoa et al., 1993). The unavoidable conclusion is that sands with ^{232}Th activity concentrations of the order of $10^6\,Bq\,kg^{-1}$ are quite likely the result of anthropogenic activities, and are thus a TENORM material. Such a conclusion is consistent with current information that the typical activity concentration of ^{232}Th in India is $3.2 \times 10^5\,Bq\,kg^{-1}$ (Pillai, 2008).

3.1.5. Radium-228

The second member of the ^{232}Th decay chain (see Table 3.2) is the radium isotope ^{228}Ra (5.75 y half-life). This radionuclide decays by beta emission into ^{228}Ac (6.15 h half-life), which in turn decays also by beta into ^{228}Th (1.91 y half-life). This latter thorium isotope decays by alpha emission into ^{224}Ra (3.66 d half-life), which decays also by alpha emission into ^{220}Rn (55.6 s half-life), a noble gas, also known as thoron, or emanation (see Table 3.2). Following the alpha decay of ^{220}Rn into ^{216}Po (0.145 s half-life), which also decays by alpha emission into the next member of the ^{232}Th decay chain, which is ^{212}Pb (10.64 h half-life). This latter radionuclide beta decays into ^{212}Bi (1.01 h half-life), which decays with a branching ratio – 64% of the time by beta emission into ^{212}Po (0.300 ms half-life) plus 36% by alpha emission into ^{208}Tl (3.05 min half-life). Both radionuclides produced by the branching process decay into ^{208}Pb (stable) which is thus the last member of the ^{232}Th decay chain; ^{212}Po decays by alpha and ^{208}Tl decays by beta (see Table 3.2). There are many gamma lines in the ^{232}Th decay chain, but the most conspicuous are the 1.615 Mev from ^{208}Tl and the ^{228}Ac photo-peaks.

It has been widely observed that the ratio $^{228}Ra/^{226}Ra$ both in soil and water is usually 1 (NAS-NRC, 1999). However, the monazite cycle may leave behind remnants containing NORM with combined ^{226}Ra plus ^{228}Ra activity concentrations that reach levels as high as 10^4 kBq kg^{-1} (see e.g., Paschoa, 2009). These NORM residues from the monazite cycle result from grinding and chemical processes, which make these radium isotopes more bioavailable than most residues resulting from non-chemical processes. Most NORM residues from the monazite cycle are found in the form of cake II (Thorium and Uranium oxides), and mesothorium (Ba, Ra SO_4). Typical ^{226}Ra plus ^{228}Ra activity concentrations in cake II and mesothorium are in the orders of 10^3 and 10^4 kBq kg^{-1}.

Several attempts have been made to solve the problems resulting from NORM residues, but to no avail (Briquet et al., 2004; Lauria and Rochedo, 2005; Pillai, 2008). One of the problems is, of course, the bioavailability of the NORM residues, which in the long run tend to reach the food chain. One additional problem is the long half-life of ^{226}Ra (1.6×10^3 y), which favors the contamination of surrounding soils and aquifers.

There are reports of measured external radiation fields of the order of only 1 μGy h^{-1} or less at the outer surface of drums containing remnants of the monazite cycle. Such low external radiation fields were attained after the original remnants were mixed with radioactive-free sand. Initially (i.e., before any deactivation or dilution in non-radioactive sand) external radiation levels at the outer walls of the cake II ranged from 10^3 to 10^4 μGy h^{-1}. Those reports on low external radiation levels take into consideration the ^{228}Ra (6.7 y half-life) and the short-term shielding effect of non-radioactive sand. However, in the long term, ^{226}Ra tends to migrate to the surfaces of the storage matrix. This fact is usually ignored, because the concentration of this radionuclide from the ^{238}U series in NORM wastes from the monazite cycle is usually one order of magnitude lower than that of ^{228}Ra. Figure 3.1 illustrates the decay and ingrowth curves of the activity ratios $^{228}Th/^{226}Ra$ and $^{228}Ra/^{226}Ra$, respectively, assuming an initial $^{228}Th/^{226}Ra$ equal to 1. One added problem, not shown in Figure 3.1, is that sometimes residual amounts of ^{232}Th stay in cake II and/or mesothorium matrices. When this is the case, the $^{228}Th/^{226}Ra$ and $^{228}Ra/^{226}Ra$ ratios do not decay as shown in Figure 3.1, but stay almost constant (horizontal) because of the continuous input of ^{228}Ra due to the alpha decay of ^{232}Th with a half-life of 1.41×10^{10} y.

At the beginning, the cake II drums are sealed but the integrity of the drums does not stand the test of time, because the ^{226}Ra activity will be there long after the lifetime of the drums, which are seldom kept under adequate conditions of a well built and long-term managed waste repository. Short-term shielding will not solve the long-term problem posed by the bioavailability of ^{226}Ra in cake II. In the long run, a better solution ought to be found to contain and store the monazite cycle remnants.

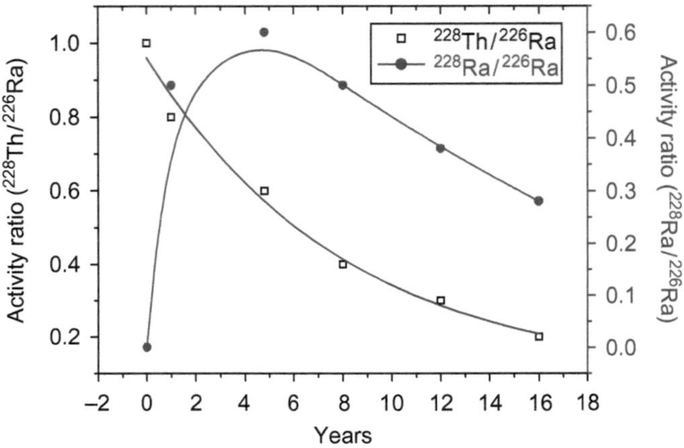

Figure 3.1 Graphical representation of the decay and ingrowth curves of the activity ratios ^{228}Th/^{226}Ra and ^{228}Ra/^{226}Ra, respectively. Figure taken from Paschoa (2009).

As one can see, ^{228}Ra is not *per se* an unmanageable residue problem, but it is closely associated with the isotope ^{226}Ra, and it may be used as an indicator of future management problems. The easily detected gamma-ray line at 911 keV of ^{228}Ac constitutes a helpful indicator of ^{228}Ra presence in NORM-contaminated soil. When the ^{228}Ra/^{226}Ra ratio is close to 1, or at least not too high, the determination of ^{228}Ra may serve as an indication of the ^{226}Ra in a contaminated soil. This procedure has already been suggested for use in identifying contaminated soil from oil fields, and successfully adopted (IAEA, 2003a, 2003b; Al-Masri and Aba, 2007).

3.1.6. Thoron (^{220}Rn)

Contrary to the major attention given to ^{222}Rn (radon) over recent decades, ^{220}Rn (55.6 s half-life) has received much less attention. However, particularly in the last decade, the number of investigations on thoron has been increasing. Thoron (see Table 3.2) is the 6th member of the ^{232}Th chain, being the result of the alpha decay of ^{224}Ra (3.66 days half-life). It is worth mentioning that a study carried out in China indicated that the ratio of the average equilibrium equivalent concentration of thoron progeny (EEC_{Tn}) to that of radon progeny (EEC_{Rn}) – EEC_{Tn}/EEC_{Rn} – ranged from 0.06 to 0.13, but the range of the dose ratio (Tn progeny)/(Rn progeny) was between 0.31 and 0.47 (Guo et al., 2005). This ratio indicates that the dose from Tn progeny may be almost half that from Rn progeny.

Moreover, the range of the EEC_{Tn}/EEC_{Rn} indoors worldwide is from 0.01 to 0.5 (UNSCEAR, 2000). The dose from inhalation of thoron and radon may be estimated by

$$E_{\text{Tn or Rn}} = \text{EEC}_{\text{Tn or Rn}} f_{\text{Tn or Rn}} T \tag{3.6}$$

where $f_{\mathrm{Tn\ or\ Rn}}$ – in $\mathrm{nSv(Bq\,h\,m^{-3})^{-1}}$ – is 40 for thoron and 9 for radon, and T the exposure time indoors is 77% of a year, or 6.72×10^3 h.

Using Equation (3.6), the estimated dose ratio (Tn progeny)/(Rn progeny) could range from about 4.4×10^{-2} to 2.2. Although the values reported by UNSCEAR cannot substitute site-specific measurements, they indicate that in some cases the indoors dose from Tn progeny may be higher than that from Rn progeny.

The determination of the thoron flux near gas mantle lamps and lanterns may be gaining momentum because the doses incurred by readers who use this kind of illumination are not yet well established. Recently, there was an attempt to control the thoron emanated from lantern lamps (Sorimachi et al., 2009).

Simultaneous measurements of radon and thoron at the same site have been improving lately (see e.g., Cosma et al., 2005; Choncubhair et al., 2008; Tommasino et al., 2008; Tokonami et al., 2008a; Ishikawa et al., 2008; Hosoda et al., 2008; Oka et al., 2008). In addition, the Japanese radon and thoron reference chambers described recently (Tokonami et al., 2008b; Sorimachi et al., 2008) have started being used elsewhere to calibrate thoron detectors (Tommasino et al., 2008). Quite recently, the Physikalisch Technische Bundesanstalt (PTB) started operating a ^{220}Rn (thoron) progeny chamber in which ^{222}Rn (radon) progeny are used for calibrating systems under realistic climatic conditions (Röttger et al., 2009).

It is interesting to note that 18% of the total papers presented at the 8th International Symposium on the Natural Radiation Environment (NRE-VIII) held in Búzios, Rio de Janeiro, Brazil, October 7–12, 2007 were radon plus thoron related. However, the ratio (number of papers on ^{220}Rn – thoron)/(number of papers on ^{222}Rn – radon) increased from 1/7 at the NRE-VII (7th International Symposium on the Natural Radiation Environment held in Rhodes, Greece, May 20–24, 2002) to 1/2 at the NRE-VIII. It is expected this kind of ratio will be even higher at the forthcoming NRE-IX Symposium.

3.2. TENR Industries

3.2.1. Oil and gas industry

The presence of radioactivity in crude petroleum was first detected at beginning of the 20th century (Burton, 1904). Soon after, it was suggested that the high radium concentrations found at the bottom of the ocean resulted from the continuous supply of radium-enriched organic materials from above (Joly, 1908). Later on, this hypothesis was confirmed by experimental research (Pettersson, 1930; Piggot, 1933).

About 20 years after Joly presented his suggestion, a new hypothesis was proposed by Samuel C. Lind proposing that the energy of alpha radioactivity might play a role in petroleum origin (Lind and Bardwell, 1926; Lind, 1931). Such a hypothesis was discussed by many investigators before and at the beginning of WWII using a variety of scientific arguments (Goodman et al., 1939; Bell et al., 1940). Almost from the beginning there was strong opposition to Lind's hypothesis (Brooks, 1931). However, in 1944 Sheppard suggested that 2.5% organic matter in a uniform, homogeneous sediment matrix containing 2.4×10^{-4} uranium per gram (i.e., a little more than 80 times the average ^{238}U abundance in the Earth's continental crust, which is $\sim 3\,Bq\ ^{238}U\,g^{-1}$) plus 8×10^{-4} thorium per gram (i.e., a little more than eight times the average ^{232}Th abundance in the Earth's crust, which is $\sim 3\,Bq\ ^{232}Th\,g^{-1}$), after about 10^8 years, would result in 40% of the total organic matter in the matrix being converted into petroleum (Sheppard, 1944). Sheppard did propose a simple differential equation based on the then current knowledge to describe the conversion rate of organic matter into petroleum (Sheppard, 1944).

On the contrary, in 1948 Brooks stated that petroleum could not have been formed by radioactivity in the organic-matter-bearing sediments because of the following (Brooks, 1948): "(a) the composition of natural gas, the complete absence of hydrogen and only the occasional presence of helium; (b) there is no relation between the abundance of petroleum and natural gas and the proximity of radioactive minerals (intensity of alpha radiation); (c) the absence of olefinic unsaturated hydrocarbons in gases of petroleums; (d) optically active hydrocarbons in petroleum could not be expected, but petroleums do contain slightly optically active hydrocarbons; (e) substances found in petroleums and derived from organic matter buried in the sediments, such as chlorophyll, porphyrins, naphtalenic acid complexes, nitrogen bases, and labile sulfur complexes, must either escape destruction by alpha radiation or those substances must be picked up subsequently by the oil, possibly during migration. "In addition, Brooks cited Sheppard who had written in his article the following (Sheppard, 1944): "In this survey of the origin of petroleum from the physicist's point of view, no conclusions can be drawn as to whether or not radioactive processes are significant in petroleum genesis."

A small group at MIT consisting of geologists, physicists, and chemists started in 1942 to examine the effect of radioactivity on the transformation of marine organism materials into petroleum hydrocarbons (Shrock, 1982). Whitehead examined evidence gathered by the MIT group from 1943 to 1952 and recognized that the amounts of petroleum found in sedimentary rocks were much too large to be explained by the alpha radioactivity of the ^{238}U and ^{232}Th and progenies found in sedimentary rocks associated with petroleum (Whitehead, 1954). It is worth mentioning that, at the end of the 1930s, all members of the MIT group were forced to turn their attention to the war effort (Shrock, 1982). It was not until after 1944 that a renewed group at MIT restarted work on the petroleum origin project. Shrock's book and the 40 articles and reports produced at MIT under the American Petroleum Institute (API) sponsored Project 43C are worth reading for a better understanding of the progress and eventual discontinuation of the project. In 1954, Whitehead presented a summary of the MIT group findings in a symposium on Nuclear Geology (Whitehead, 1954). For a long time thereafter, there were no more articles published in the open literature associating natural alpha radioactivity and petroleum genesis. The one exception was a phrase in a classical textbook on geochemistry (Krauskopf, 1967) mentioning that bombardment of fatty-acid molecules with alpha particles emitted by naturally occurring radionuclides could be a possible mechanism in producing hydrocarbons.

The uranium contents of some oil field brines have been reported to be 2.4 Bq ^{238}U kg^{-1} in association with helium-rich petroleum and natural gas (Pierce et al., 1955). The relationship of helium to oil and to petroleum hydrocarbons has become evident since the 1950s (Anderson and Hinson, 1951; Pierce et al., 1955; Boone, 1958; Nicanov, 1969; Tverdova and Fedina, 1974). In fact, the alpha radiation theory of petroleum genesis predicts that helium can be expected to accumulate in small amounts during the formation of petroleum and natural gas as a result of the alpha decay of uranium and thorium and their respective progenies.

Until the mid-1970s there was no evidence that organic-rich sediments, like sapropels, might have high concentrations of uranium and thorium (see e.g., Fischer and Arthur, 1977). In addition, some petroleum geologists stated at that time that sapropelic kerogen related to subaqueous sediment deposits was associated with rich oil sources under conditions of isolation from land areas (Markovskii, 1978; Tissot and Welte, 1978). Several sapropelic episodes might have resulted in black shales with uranium concentrations higher than those of sedimentary rocks (see Getseva, 1958; Wedepohl, 1978). Moreover, it has been reported that a sapropel found in the Black Sea dating from about 5k years had uranium concentrations of about 40 ppm ($\sim$0.5 Bq^{238}U g^{-1}) mostly associated with plankton rather than land-derived organic-matter debris (Degens et al., 1977). In addition, the form of occurrence of uranium in petroleum deposits associated with organic matter suggested once more that uranium-bearing substances were genetically associated with petroleum (Vassiliou, 1980).

The above, plus further information that became available in the open literature just prior to 1980, made the objections raised by Brooks decades before (Brooks, 1948) all but meaningless. Thus, in 1981, a paper presented at an international symposium on the natural radiation environment suggested that the alpha radiation theory of petroleum origin needed to be reexamined in the light of the new evidence, in particular the presence of ^{238}U and ^{232}Th plus their respective progenies present in sapropel (Paschoa, 1981).

This paper was unnoticed for more than a decade, though the problem of NORM in the oil and gas industry had already been recognized by a number of investigators (Gesell, 1975; Perel'man, 1977). As a matter of fact, radioactivity in oil waters and in oil well samples had been observed a long time back: (i) when Vernadsky called attention to the presence of ^{226}Ra and ^{228}Ra in brines of oil deposits in Russia (Vernadsky, 1930; Vernadsky and Chlopin, 1932; Bäshounek et al., 1935); and (ii) later when Bruno Pontecorvo suggested the use of radioactivity in these samples to correlate with the results on the radioactivity of well logs (Pontecorvo, 1942).

In the late 1980s and early 1990s, the NORM problem in the oil and gas industry had become an annoyance to the extent that the API proposed options for disposal of NORM wastes produced in the American petroleum industry (API, 1988, 1992). The radioactive scales growing in oilfield pipelines and tubing have been known for many years. Concentrations of ^{226}Ra reached levels up to $1 \times 10^3\,\mathrm{Bq\,g^{-1}}$ (Kolb and Wojcik, 1985; Steinhäusler, 1985; Heaton and Lambley, 1995; Kvasnicka, 1996).

The production processes of the oil and gas industry mobilize natural radionuclides extant in the reservoirs from which these materials are extracted. As a consequence, some natural radionuclides are deposited as scales in transporting tubes, sludge in tanks, pumps, and a variety of other equipment. NORM appears also after the separation of the oil and gas phases, and in the remnants of the production water. A wide variation of activity concentrations of ^{238}U and ^{232}Th plus radionuclides of their respective progenies, mainly ^{226}Ra, ^{222}Rn, and ^{210}Pb (from the ^{238}U series), and ^{228}Ra, ^{228}Th (from the ^{232}Th series) appears in scales and sludge.

The association between the occurrence of NORM in the oil and gas industry and its implication concerning petroleum genesis escaped the attention of most investigators, with a sole exception (Paschoa, 1997). This author stated that the geological ages accepted for the formation of most known petroleum reserves are within a time span that makes natural alpha radiation energy a strong candidate as one of the energy sources for the production of petroleum hydrocarbons. In addition, he suggested that future research on the relationship between NORM and petroleum genesis should also include the following:

1. systematic studies on ^{238}U and ^{232}Th concentrations in sapropel and other organic materials known to be associated with petroleum genesis;
2. knowledge of the microdistribution of long-lived natural alpha emitters in the organic segment of sapropel;
3. determination of the fractions of radon and thoron that escape from the organic part of the sediments;
4. determination of the mean diameter of the organic constituents of sapropel;

5. correlations between the age of the petroleum and the concentrations of ^{238}U and ^{232}Th; and
6. comparisons between the $^{238}U/^{232}Th$ and $^{226}Ra/^{228}Ra$ ratios in sapropel and NORM residues of the oil industry.

It is well known that ^{238}U and ^{232}Th are largely immobile in rocks whether the rocks are associated with petroleum or not. On the contrary, radium is more soluble and may enter some fluid phases of the oil formation. Terrestrial radionuclides like ^{222}Rn, ^{210}Pb, and ^{210}Po from the ^{238}U series, or ^{228}Th, ^{224}Ra, ^{220}Rn, and ^{212}Pb are of lesser importance from the radiological viewpoint in the oil industry.

Table 3.10 summarizes selected data found in the open literature on radon concentrations for different phases of the oil and gas industry. As mentioned in Section 3.1.3 (on radon), indoor ^{222}Rn activity concentrations throughout the world (as reported by UNSCEAR, 2000) range from 24 Bq m^{-3} in Egypt (Kenaway and Morsy, 1991) to about 8.5×10^4 Bq m^{-3} (Swedjemark et al., 1993). Comparing this range with the range of all data presented in Table 3.10, one can note that the two highest values – 1.8×10^4 Bq m^{-3} (reported in an oil shale) and 2×10^5 Bq m^{-3} (reported in natural gas) – are about the same order of magnitude as the highest indoor radon concentration. It is clear, however, that the high radon concentration reported for some Swedish houses (8.5×10^4 Bq m^{-3}) triggers actions based on some of the fundamental safety principles such as those defined by IAEA (2006):

1. principle 4 (justification of facilities and activities);
2. principle 5 (optimization of protection);

Table 3.10 Radon concentrations in selected phases of the oil and gas industry.

Produced water	Radon concentration (Bq m^{-3})			
	Crude oil	Natural gas liquids	Oil shale	Reference(s)
–	–	5×10^2	–	Kolb and Wojcik (1985)
–	340	–	–	UNSCEAR (2000)
–	–	$5–2.0 \times 10^5$	–	IAEA (2003b)
9.8×10^{-4}– 1.9×10^{-2}	2×10^{-2}–0.3[a]	3.0×10^2– 2.5×10^3		Hamlat et al. (2003)
–	–	–	1.8×10^4	
–	–	$15–1.1 \times 10^3$	–	Al-Masri and Shwiekani (2008)

[a] Bq g^{-1}

3. principle 6 (limitation of risks to individuals);
4. principle 7 (protection of present and future generations); and
5. principle 10 (protective action to reduce existing or unregulated radiation risks).

The IAEA published a few years ago a Safety Report Series to be used by the oil and gas industry with respect to radiation protection and the management of radioactive waste (IAEA, 2003b). This IAEA publication has a chapter that deals specifically with NORM in the oil and gas industry.

The oil and gas industry operates throughout the world in different climates and over a wide range of latitudes and longitudes. NORM in the oil and gas industry constitutes a common problem for this industry in many countries. At the same time, the differences in culture, environment, industrial practice, and legislation make NORM in the oil and gas industry a unique opportunity for international cooperation, and a tremendous challenge.

Variable concentrations of ^{238}U and ^{232}Th plus those radionuclides from their respective progenies – mostly ^{226}Ra, ^{222}Rn, ^{210}Po, and ^{210}Pb from the ^{238}U series, and ^{228}Ra, ^{228}Th, ^{224}Ra, ^{220}Rn, and ^{212}Pb from the ^{232}Th series – are present over a wide range of concentrations in formation water, scales in the interior walls of pipes and equipment, sludge in separation and production equipment, and residues stored in drums. Table 3.11 shows typical NORM (excluding radon) activity concentration levels for the oil and gas industry in these materials. Extant natural radionuclides are mobilized in the oil reservoirs during the extraction processes. As the natural pressure of an oil reservoir drops, formation water is added to bring the oil to the surface. Coprecipitation of radium occurs when barium sulfate ($BaSO_4$), calcium sulfate ($CaSO_4$), and calcium carbonate ($CaCO_3$) are added to the production water. Radium concentrations may increase as the salinity of produced water also increases. A factor of 60 increase is known to occur in

Table 3.11 Typical activity concentration values in NORM-bearing scales, sludge and films from the oil and gas industry.

Material	Typical range or value of activity concentration
Formation water	1.9×10^3–1.11×10^5 Bq ^{226}Ra m^{-3}
Scales in pipes and equipment	10^3–10^5 Bq ^{226}Ra kg^{-1}
Sludge in separation and production equipment	10^4–10^6 Bq ^{210}Pb kg^{-1}
Sludge and film in natural gas equipment	4.0×10^9 Bq ^{226}Ra m^{-3}
Residues stored in drums	$\geq 3.3 \times 10^3$ Bq kg^{-1}

Source: Data from API (1996), Seitz (1996).

the salinities of production water. Because radium coprecipitates with alkaline earth elements (i.e., calcium, strontium, and barium) – all four in Group IIA – to form carbonates, sulfates, and silicates, radium in these chemical forms may be found in production water, scales, and equipment.

Much is already known on how to deal with radioactivity in general and with natural radioactivity in particular. Those with knowledge of radiation protection are convinced that the NORM problem is tractable, once the subject is adequately regulated by the oil industry and national authorities. Regulation, but not overregulation, is necessary to deal with this problem.

Typical average activity concentrations in scales and sludge in the United States oil and gas industry are presented in Table 3.12. In gas plants the most significant natural radionuclides present are ^{210}Po and ^{210}Pb, which originate from the decay of ^{222}Rn.

It can be observed from Table 3.12 that the average activity concentrations reported by the US EPA are for all but one (^{210}Po in scales) within the ranges of the respective radionuclide activity concentrations in scales and sludge found elsewhere in the world. Within the worldwide range one can find, for example, ^{226}Ra and ^{228}Ra activity concentrations equal to 1.1×10^5 and 1.3×10^5 Bq kg^{-1}, respectively, found in scales of the oil industry in Malaysia (Omar et al., 2004).

The current efforts being made by the IAEA, the European Commission and other international boards are first but important steps toward solving the NORM problem in most parts of the world. The awareness by the oil industry of the NORM problem, as well as good

Table 3.12 Typical average and range of activity concentrations in scales and sludge in the oil and gas industry.

	Bq kg^{-1}			
	Scales		Sludge	
	USEPA	Jonkers et al. (1997)[a]	USEPA	Jonkers et al. (1997)[a]
^{238}U	–	$1.0–5.0 \times 10^2$	–	5.0–10.0
^{226}Ra	1.3×10^4	$1.0 \times 10^2–1.5 \times 10^7$	2.1×10^3	$50–8.0 \times 10^5$
^{210}Pb	1.3×10^4	$20–7.5 \times 10^4$	2.1×10^3	$1.0 \times 10^2–1.3 \times 10^6$
^{210}Po	1.3×10^4	$20–1.5 \times 10^3$	2.1×10^3	$4.0–1.6 \times 10^5$
^{232}Th	–	1.0–2.0	–	2.0–10
^{228}Ra	4.4×10^4	$50–2.8 \times 10^6$	7.0×10^2	$5.0 \times 10^2–5.0 \times 10^4$
^{228}Th	4.4×10^4	–	7.0×10^2	–

Source: Adapted from USEPA (1993), Jonkers et al. (1997), and IAEA (2003b).
[a]Cited in IAEA (2003b).

radiation hygiene practices, may make the difference in bringing this problem into its actual perspective and under control.

Table 3.13 summarizes the data on the production of crude oil and natural gas for the main oil and gas producers in the world. The data presented in Table 3.13 indicate that countries as different as Russia, Saudi Arabia, the United States, the Islamic Republic of Iran, and China are responsible for the production of more than 40% of the world production of crude oil and natural gas. The diversity of the radiation protection cultures in the main producers of crude oil and natural gas indicates clearly the need for international guidance in dealing with the NORM problem, as recommended by the IAEA (2003b).

The wide range of ^{226}Ra and ^{228}Ra activity concentrations found in scales and sludge resulting from E&P operations is a common problem to be solved by the oil and gas industry. Some attempts have met with different degrees of

Table 3.13 Crude oil and natural gas production in 2007 by selected producers, and the global production.

Producer	Crude oil[a] (million barrels per day) (%)	Natural gas[b] (billion m^3) (%)
Russia	9.44–12.9	654–21.6
Saudi Arabia	8.72–11.9	76–2.5
USA	5.10–7.0	546–18.1
Islamic Republic of Iran	3.92–5.4	112–3.7
China	3.73–5.1	69–2.3
Mexico	3.08–3.6	56–1.9
Canada	2.61–3.6	187–6.2
United Arab Emirates	2.60–3.5	49–1.6
Kuwait	2.46–3.4	13–0.004
Venezuela	2.43–3.3	27–0.009
Nigeria	2.35–3.2	34–0.011
Norway	2.27–3.1	93–0.030
Brazil	2.27–3.1	10–0.003
Iraq	2.09–2.9	4–0.001
United Kingdom	1.50–2.0	72–0.024
Subtotal	54.57–74.5	2,002–66.3
Other nations	18.70–25.5	1019–33.7
Global total production	73.27–100.0	3,021–100.0

Source: Based on data from EIA (2008) and CIA (2009).
[a]EIA, Energy Information Agency (USA).
[b]CIA, Central Intelligence Agency (USA).

success. A concerted international effort will be necessary to deal properly with the diffuse NORM produced by the oil and gas industry throughout the world.

One current proposal to solve the problem posed by the large volume of NORM wastes (non-biologically available) from the oil and gas industry is, for example, to decrease the activity concentration of NORM wastes (biologically available) from the mineral sands industry (Paschoa, 2009).

3.2.1.1. Agreements relevant to dumping radioactive wastes into the sea

At this point, it is interesting to mention that the two main international agreements concerning the dumping of radioactive wastes into the seas and oceans are, from 1972, the Convention on the Prevention of Marine Pollution by Dumping of Wastes and other Matters, better known as the London Convention, and, from 1977, the Multilateral Consultation and Surveillance Mechanism for the Sea Dumping of Radioactive Wastes, which was established under the auspices of Organization for Economic Co-operation and Development (OECD) (IAEA, 1975, 1978a, 1978b, 1978c). It is also worth mentioning that the original version of the London Convention prohibited, *inter alia*, dumping of high-level wastes into the sea. The low-level radioactive wastes, however, were allowed to be dumped into the sea under special permits. The IAEA became responsible, under the London Convention, for defining what kind of high-level radioactive waste was unsuitable for dumping into the seas and oceans. As a consequence, IAEA made a series of recommendations for dumping of other radioactive wastes and for keeping the subject under permanent review. The strong opposition by world public opinion induced the London Convention to prohibit the dumping of all kinds of radioactive wastes into the sea, which came into effect in 1993 (London Convention, 1993). Moreover, a 1996 Protocol intended to substitute the 1972 Convention and to be much more restrictive also introduced two interesting principles: (i) Article 3 – the "polluter pays principle"; and (ii) prohibition of the dumping of any wastes or other matter with the exception of the following: dredged material; sewage sludge; fish waste, or material resulting from industrial fish processing operations; vessels and platforms or other man-made structures at sea; inert, inorganic geological material; organic material of natural origin; bulky items primarily comprising iron, steel, concrete and similar unharmful materials for which the concern is physical impact and limited to those circumstances, where such wastes are generated at locations such as small islands with isolated communities, having no practical access to disposal options other than dumping. The exceptions were cases of *force majeure* caused by stress of weather, or in any case that constitutes a danger to human life or a real threat to vessels. The 31st Meeting of the Scientific Group under the London Convention and the 2nd Meeting of the Scientific Group under the London Protocol occurred in Guayaquil, Ecuador, in May 19–23, 2008.

The MARINA Project was launched in 1985 (thus before Chernobyl) to assess the radiological exposure of the European Community from radioactivity in North European waters (CEC, 1989, 1990). The proposed Environmental Impact Assessment (EIA) proposed at the time comprised the evaluation of the following:

(i) the sources and pathways of radioactive substances and their concentrations in the maritime area;
(ii) the radiation exposure of humans and marine ecosystems;
(iii) biological and ecological effects in the marine environment in their own right, including the vulnerability of marine ecosystems, arising from existing and future foreseen discharges, emissions and losses of radioactive substances However, such data were rather scarce and developments concerning impact assessment were still in a preliminary stage; and
(iv) other adverse effects that may affect other legitimate uses of the sea

An important outcome of a meeting held in Paris (September 21–22, 1992) of the so-called Oslo and Paris Commissions (OSPAR) was the Convention for the Protection of the Marine Environment of the North-East Atlantic, better known as the OSPAR Convention (OSPAR, 1995). There is today an OSPAR site on the internet, http://www.ospar.org, which states that its "mission is to conserve marine ecosystems and safeguard human health in the North-East Atlantic by preventing and eliminating pollution; by protecting the marine environment from the adverse effects of human activities; and by contributing to the sustainable use of the seas."

In 1996, all the data available on marine radioactivity in seawater, sediments, and biota were stored by the IAEA Marine Environment Laboratory (IAEA-MEL) as the Global Marine Radioactivity Database (GLOMARD) (Povinec et al., 1996a). At just about the same time, the IAEA-MEL organized studies on the sources of radioactivity in the marine environment, as well as on their relative contributions to overall dose assessment, which became known as the MARDOS Project (Baxter et al., 1996; Aarkrog et al, 1997). The MARDOS Project provided estimates of doses to the public from ^{137}Cs from global fallout, the Chernobyl accident, authorized discharges and natural ^{210}Po through consumption of marine food (Povinec et al., 1996b).

The IAEA-MEL also coordinated the Worldwide Marine Radioactivity Studies (WOMARS) project to achieve the following objectives (Povinec et al., 2005): (i) to identify the major sources of anthropogenic radionuclides in the world ocean; (ii) to develop present knowledge of the distributions of key radionuclides (^{90}Sr, ^{137}Cs, and Pu isotopes) in water and sediment of the world ocean; and (iii) to study the evolution of radionuclide concentrations in water with time using quality controlled data (e.g., from the GEOSECS (Geochemical Ocean Sections Study) program of the mid-1970s (Bowen et al., 1980)), and new data sets.

Despite all the efforts made in recent decades, it is recognized today that knowledge of oceanic processes for development of a comprehensive model to describe the behavior of radionuclides extant or released into the seas and oceans is not adequate. The IAEA, however, has some time ago proposed principles for establishing limits for the release of radioactive materials into the environment (IAEA, 1978a, 1978d, 1985). The approach adopted by IAEA in issuing such principles was to evaluate the environmental capacity and establish limits within which the quantities of radioactive waste could be controlled (i.e., the critical pathways approach) (IAEA, 1978b, 2001). The derived limits for releasing radionuclides into the environment were based on the ICRP Publication 29 (ICRP, 1979). Release-rate limits were derived for each particular radionuclide and for a number of pathways. A selection was made in order to include those pathways known to exist at the time of derivation plus some others, which might become important in the foreseeable future.

The release-rate limits for alpha emitters for a single site were of the order of 10^{15} Bq y^{-1}, but limited to 10^{14} Bq y^{-1} for ^{226}Ra and supported ^{210}Po (IAEA, 1978c). However, to meet the earlier requirements of the London Convention, it was necessary to express the derived limits in terms of concentration, which meant about 10^{10} Bq mt^{-1} for alpha emitters, but 10^{9} Bq mt^{-1} for ^{226}Ra and supported ^{210}Po, assuming an upper limit to the mass dumping rate of 10^{5} mt y^{-1} at a single dumping site (IAEA, 1978a; Templeton, 1981).

The levels of radioactivity in the marine environment had already led some investigators in the 1970s and early 1980s to pay attention to the biological availability of selected chemical forms of radionuclides, as well as the resulting internal doses to marine organisms (Templeton et al., 1971; Woodhead, 1973; Paschoa and Baptista, 1978; Paschoa et al., 1979, 1981b). At that time, it was concluded that in the long term ^{137}Cs would be the major contributor to the total dose rate for all groups of organisms, but in the case of phytoplankton ^{239}Pu could be of some importance (Woodhead, 1973). Taking into account the data gathered in the 1970s and 1980s, it is reasonable to hypothesize that alpha emitters from NORM may be of some importance for the alpha radiation dose to aquatic biota, mostly because of the high bioaccumulation factors of some natural emitters for plankton, algae, invertebrates, and macrophytes.

Early in the 21st century, IAEA launched the EMRAS program at a meeting held in Vienna, September 1–5, 2003. In subsequent years, there were four more combined meetings with a larger number of participants from a greater number of countries, as illustrated in Figure 3.2. After a steep increase from 2004 (2nd combined meeting) to 2005 (3rd combined meeting) in the number of participants and participating countries there was

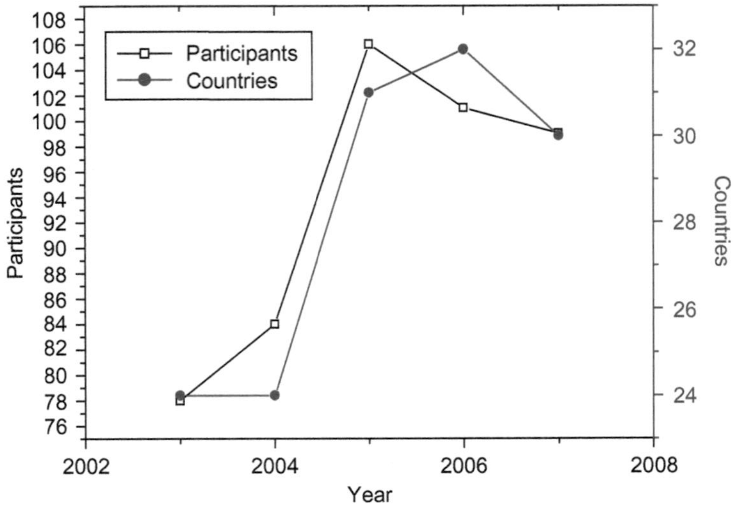

Figure 3.2 Number of participants and countries in five successive EMRAS meetings.

a decrease from 2005 to 2007 (5th combined meeting). However, the seven established Working Groups continued to operate. The Working Group on Modelling of NORM, for example, agreed that the information available was not in a form to be used promptly by the program. Information is in the process of being collected from NORM industries on the available data and existing models. The fact is that, despite several meetings, some working groups, for example, that on Modelling of NORM, did not achieve a desirable result under EMRAS. As a consequence, the EMRAS II program was launched at the IAEA's Headquarters in Vienna, in January 19–23, 2009. EMRAS II will continue the effort on radiological modelling, which started in previous programs. Among other tasks, Working Group 2 of the EMRAS II program intends to develop: (i) a handbook of transfer parameter values for application to biota and (ii) methods for determining protection quantities and levels for biota based on the existing data available on radiobiological effects. In addition, it was suggested that the working group should interact with international activities, like those of ICRP and EC, as deemed necessary.

3.2.1.2. Bioaccumulation of alpha emitters

A database of bioaccumulation (i.e., relative concentrations) for selected alpha emitters, such as $^{239,240}Pu$, ^{238}U, and ^{226}Ra, in aquatic biota was summarized years ago by the National Council on Radiation Protection (NCRP) from the United States (NCRP, 1984). It is well known, however, that there are a number of uncertainties in the determination of bioaccumulation factors – assumed here to be *the radionuclide concentration, at equilibrium, in the organism divided by the concentration of the radionuclide in the water* – for any radionuclide mostly due to conversion factors for wet weight and dry weight, which vary from species to species. As shown in Table 3.18, reported values for bioaccumulation factors for aquatic biota indicate ranges over several orders of magnitude.

One can state, taking into account the orders of magnitude presented in Table 3.18, that phytoplankton and zooplankton, algae, invertebrates, and macrophytes may receive potentially high alpha dose rates. For the sole purpose of illustrating the potential alpha dose rates for highly bioaccumulating organisms, the discussion that follows will be restricted to phytoplankton and zooplankton. Here one must keep in mind that in some cases part of the alpha energy emitted internally in plankton may be dissipated in the surrounded water (Woodhead, 1973). On the contrary, for many species of plankton, the ranges of the alpha particles emitted internally by natural radionuclides and artificially produced actinides are of the same order as the linear dimensions of the organisms (Paschoa et al., 1979, 1981).

Further research is needed to improve knowledge on the fraction of the energy of internally emitted alpha particles that escapes the organism volume. Such knowledge has tantamount importance in the dosimetry of aquatic organisms with high bioaccumulation factors for alpha emitters.

3.2.1.3. Alpha dose to plankton

It is well known that radionuclides produced in the fuel cycle enter the aquatic environment by a variety of pathways. Among those radionuclides thus produced, one should include the technologically enhanced natural alpha emitters (TENRAE) in the pre-reactor phase and the transuranium alpha emitters (TAE) in the post-reactor phase of the nuclear fuel cycle. The orders of magnitude estimated for the ratios of plankton dose rates are up to 10 for fallout/natural background, and from 10^2 to 10^3 for waste disposal at Windscale/natural background (Preston, 1975). Specific comparisons, however, between internal dose rates from man-made and natural alpha emitters have rarely been reported (Woodhead, 1973; Paschoa and Baptista, 1978; Paschoa et al., 1979, 1981). Considering the bioaccumulation factors reported in Table 3.14, it is reasonable to hypothesize that similar values may occur in marine biota. Thus, in accordance with this hypothesis, one can expect that NORM concentrations in marine environments may lead to high annual internal alpha radiation doses to marine biota, particularly to plankton.

Table 3.15 presents, for illustration purpose only, the range of absorbed dose rates due to internally incorporated ^{226}Ra in plankton. Two interesting aspects are worth mentioning here concerning the relatively

Table 3.14 Ranges of bioaccumulation factors for $^{239,240}Pu$, ^{238}U, and ^{226}Ra in freshwater biota.

	Bioaccumulation factor			
	$^{239,240}Pu$		^{238}U and ^{226}Ra	
	Low	High	Low	High
Piscivorous	10^{-1}	10^{0}	10^{-1}	
Planktivorous	10^{-1}	10^{1}	10^{-1}	10^{0}
Omnivorous	10^{-1}	10^{1}	10^{-1}	10^{0}
Benthic	10^{-1}	10^{2}	10^{0}	10^{2}
Zooplankton	10^{2}		10^{2}	
Phytoplankton	10^{2}	10^{4}	10^{1}	10^{3}
Algae	10^{3}		10^{1}	10^{3}
Invertebrates	10^{2}	10^{3}	10^{0}	10^{3}
Macrophytes	10^{2}	10^{4}	10^{1}	10^{3}

Source: Adapted from NCRP (1984).

Table 3.15 Absorbed dose rates from internally incorporated ^{226}Ra in plankton.

Dose rate ($Gy\,h^{-1}$)		Reference(s)
Low	High	
6.1×10^{-4}	2.0×10^{-3}	Woodhead (1973)
2.0×10^{-4}	[a]1.6×10^{-1}	Paschoa et al. (1981)

[a]This high dose rate was estimated from ^{226}Ra incorporated in the phytoplankton *Chaetoceros* and *Rhizosolenia* from Walvis Bay in the west coast of Namibia.

high value, $1.6 \times 10^{-1}\,Gy\,h^{-1}$ of absorbed dose rate, which appears in Table 3.15. Firstly, the average ^{226}Ra activity concentration corresponding to this high dose rate can be traced back to seven phytoplankton samples taken from 15 plankton samples collected in the Walvis Bay area (Shannon and Cherry, 1971). Secondly, this high $1.6 \times 10^{-1}\,Gy\,h^{-1}$ absorbed dose rate from ^{226}Ra is comparable to the dose rate from ^{239}Pu accumulated in the zooplankton *Gammarus* from Thule, Greenland as the result of a release associated with a military airplane accident (Paschoa and Baptista, 1978). The concentrations of ^{239}Pu in marine organisms living in the surroundings of the Thule accident, however, decreased significantly with time (Aarkrog, 1977).

To the best of my knowledge, and in accordance with historical accounts, it is quite likely that the ^{226}Ra concentrations in the Walvis Bay area had been enhanced by more than 100 years of gold mining, and then, in the second half of the 20th century, by heap-leaching uranium

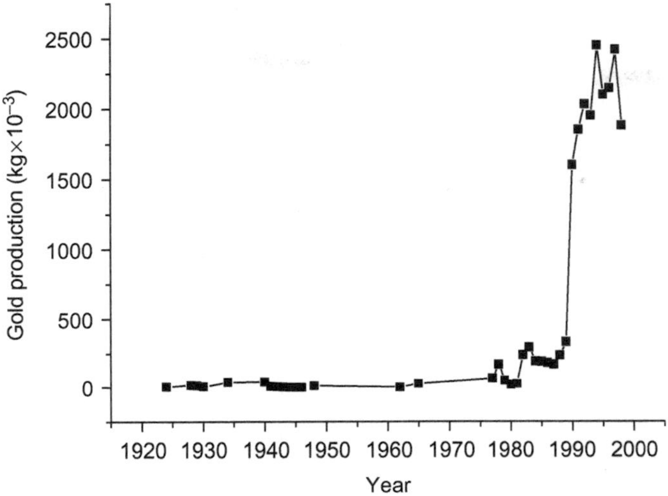

Figure 3.3 Evolution of gold production in Namibia (kg × 10^{-3}) per year from 1924 to 1998. Data taken from Burnett (1998).

extraction activities on the gold mining tailings existing in Witwatersrand and its extensions (Paschoa, 2002; see also Blight and Caldwell, 1984). Uranium is accumulated up to 0.1% (i.e., 1.0×10^3 mg kg^{-1}) in the gold tailings of Witwatersrand (Winde and De Villiers, 2002; Winde et al., 2004). Accordingly, since 1952, uranium has been produced in Namibia, which was for a long time part of South Africa, and in South Africa itself as a by-product of gold mining and extraction. Figure 3.3 represents graphically the evolution of the Namibia gold production from 1924 to 1998.

The ratio of production of U_3O_8 concentrate (mt)/gold (kg) from 1994 to 1998 is shown in Figure 3.4. This production ratio steadily decreased since 1994 because of the dismantling of South Africa's nuclear weapons program in 1993 (von Baeckmann et al., 1995).

The germane question here, however, is what percentage of the uranium series radionuclides, like ^{226}Ra, ^{210}Po, and ^{210}Pb, found in plankton off the west coast of Namibia originated in the heap-leaching gold tailings of Witwatersrand for uranium production.

3.2.1.4. Uranium and thorium series in the oil and gas industry

It is well known that in the process of producing oil and gas the natural radionuclides existing in the reservoirs are mobilized. Some of these natural radionuclides are deposited as scales in transporting tubes, and as sludge in tanks, pumps, and a myriad of other equipment. These NORM appear also

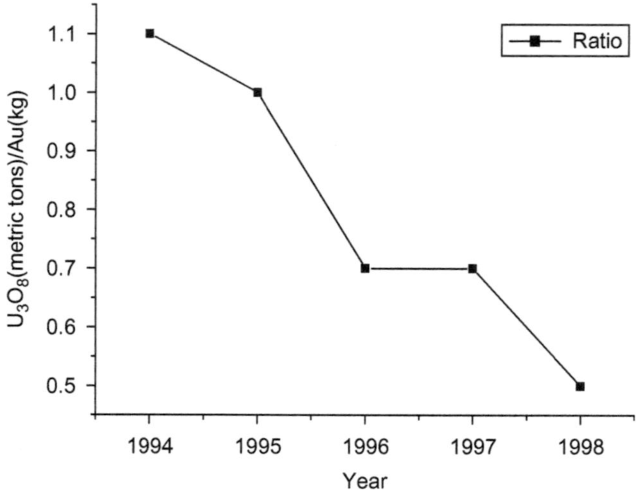

Figure 3.4 Graph of the evolution of the production ratio of [U_3O_8 concentrate (mt)/gold (kg)] in South Africa, from 1994 to 1998.

Table 3.16 Typical range or average of activity concentrations in NORM-bearing scales, sludge and films from the oil and gas industry.

Radionuclide(s)	Material	Unit	Typical range or average
	Scales in pipes	$Bq\,kg^{-1}$	10^3–10^5
	Formation water		1.9×10^3–1.11×10^5
	Sludge and film		4.0×10^9
^{210}Pb	Sludge in equipment	$Bq\,kg^{-1}$	10^4–10^6
^{222}Rn	Natural gas liquids	$Bq\,m^{-3}$	15–2.5×10^3
Miscellaneous	Residues stored	$Bq\,kg^{-1}$	$\geq 3.3 \times 10^3$

Source: Based on data from API (1996), Seitz (1996), Hamlat et al. (2003), and Al-Masri and Shwiekani (2008).

after the separation of the oil and gas phases, and in the remnants of the production water. The concentrations of radionuclides of ^{238}U and ^{232}Th series progenies – ^{226}Ra, ^{222}Rn, ^{210}Po, and ^{210}Pb from the ^{238}U series, and ^{228}Ra from the ^{232}Th series – vary considerably in the distinct phases of the oil and gas industry, as illustrated in Table 3.16.

As mentioned previously, there are oil and gas industrial operations throughout the world. Cultural differences, distinct environments, peculiar practices, and diversity in national legislation transform the concerns on the impacts of NORM from the oil and gas E&P into a tremendous challenge but at the same time provide a unique opportunity for international collaboration.

3.2.2. Coal industry

Worldwide coal is a prime source of energy. Coal mining and its use in power plants represents one of the most important industries in relation to TENR. In the United States, for example, the average uranium and thorium concentrations in mined coal (fraction of dry weight) are 2.1×10^{-2} and 9.8×10^{-3} Bq g^{-1}, respectively (Beck, 1980; Eisenbud and Gesell, 1997).

A few examples illustrate the magnitude of the TENR problem in the coal industry: (a) in the Czech Republic, coal mining production reached more than 130 Mt y^{-1} in the past and still amounts to 70 Mt y^{-1}(Moravanska and Laciok, 2004); (b) in Italy over 65% of electricity has been provided by the ENEL Group (the largest power company in the country) in 2001, accounting for almost all of the electricity generated from coal, that is, 12% of total energy produced in 12 coal-fired power plants, using nearly 1.13×10^7 t of coal (Trotti et al., 2005); (c) in the United Kingdom, 23 coal-fired power stations (operating capabilities from 230 to 3,800 MW$_e$) represent over 40% of the national electricity production (Penfold et al., 1998); (d) in Poland, waste rock resulting from 20 years of coal mining is deposited in a storage yard, filled to about 30%, covering an area of 83 ha and forming a 25 m high waste dump hill (Komosa et al., 2004); (e) in the Upper Silesian Coal Basin, the operation of 50 underground coal mines necessitates the daily removal of up to 800,000 m^3 d^{-1} of groundwater, pumped to the surface (Leopold and Michalik., 2007); (f) in coal industries in the western Balkan countries, radon exhalation rates in the waste repository of the Koscevle mine reach 0.25 Bq m^{-2} s^{-1}, which is one order of magnitude higher than the average for the world (Klerkx et al., 2007); (g) in Foz Calanda, Spain, typical 1-year coal consumption is 4.5×10^9 kg, with production of about 18% fly ash, which has average activity concentrations of 3.1×10^2 Bq ^{40}K kg^{-1}, 1.9×10^2 Bq ^{226}Ra kg^{-1}, 74 Bq ^{232}Th kg^{-1}, and 2.6×10^2 Bq ^{210}Po kg^{-1} (Mora et al., 2008); and, at the Kardia mine in Northern Greece, the average activity concentrations of natural radionuclides in 26 samples of ash are 2.5×10^2 Bq ^{40}K kg^{-1}, 25 Bq ^{235}U kg^{-1}, 5.1×10^2 Bq ^{238}U kg^{-1}, 4.1×10^2 Bq ^{226}Ra kg^{-1}, 51 Bq ^{232}Th kg^{-1}, and 49 Bq ^{228}Th kg^{-1} (Fotakis et al., 2008).

The combustion of coal produces large amounts of wastes, such as: boiler slag and bottom ash from the boiler; lagoon sludge from the ash water basin; fly ash from the electrostatic precipitator; filter press sludge and gypsum from flue gas desulphurization (FGD); and dust from the stack. The largest amount of solid waste is in the form of fly ash.

Coal is naturally radioactive, containing traces of radionuclides in the decay series of uranium–radium, thorium, and uranium–actinium, as well as from ^{40}K. Therefore, ash produced by burning coal also contains natural radioactivity. The radionuclide concentration in coal ash depends on the radionuclide concentration in the coal, the ash content of the coal, and the conditions during combustion in the boiler of the power station. The ash content of bituminous coal used in power stations is in the range of about 7–40% (typical average: 15%), that is, the activity concentration for all radionuclides in ash is typically higher by a factor of seven than in coal. Whilst the radioactive noble gas radon (^{222}Rn and ^{220}Rn) is likely to diffuse into the air, the other nuclides in the U- and Th decay series will be present in the combustion products, for example in fly ash: ^{238}U, ^{234}Th, ^{234m}Pa, ^{234}Pa, ^{234}U, ^{230}Th, ^{226}Ra, ^{218}Po, ^{218}At, ^{214}Pb, ^{214}Bi, ^{214}Po, ^{210}Pb, ^{210}Bi, ^{210}Po, ^{232}Th, ^{228}Ra, ^{228}Ac, ^{228}Th, ^{224}Ra, ^{216}Po, ^{212}Pb, ^{212}Bi, ^{212}Po, and ^{208}Tl. In addition, coal and therefore its combustion products also contain ^{40}K.

In the 1960s and 1970s, a series of studies was undertaken in the United States mostly to compare the radioactivity releases from coal-fired and nuclear power plants (Eisenbud and Petrow, 1964; Terril et al., 1967; Martin et al., 1970; Bedrosian et al., 1970; Hull, 1971; Lave and Freeburg, 1973; Hull, 1974). The radiological impacts of airborne effluents of a model 1,000 MW_e coal-fired power plant and a model 1,000 MW_e pressurized water reactor (PWR) were compared through a study by an Oak Ridge National Laboratory team of scientists (McBride et al., 1977). Individual and population doses due to airborne effluents plus contamination of soils, vegetation, water, and animals from releases of ^{226}Ra and ^{228}Ra from a coal-fired power plant in Poland indicated values 30 times higher than those related to nuclear power plants (Glowiak and Pacyna, 1980).

The high age of coal ensures secular equilibrium between the various radionuclides (i.e., all nuclides in a given decay series have equal specific activities). In addition to the age, there is an assumption of a "closed system" for secular equilibrium to be attained. The activities of man with coal have the potential of opening the system, particularly to the loss of ^{222}Rn and ^{220}Rn. In the assaying of U and Th content by counting the gamma emissions from one of the progeny, the samples are customarily placed in sealed containers (e.g., ice cream containers sealed with tape may be used) and not counted until secular equilibrium is again obtained after 5–10 half-lives, depending on the error requirements. Therefore, it is general practice to quantify the activity concentration of representative radionuclides for the uranium and thorium decay series, such as radium (^{226}Ra) and thorium (^{232}Th), respectively.

The IAEA-MEL has documented how non-nuclear industries, like the coal industry, technologically enhanced the natural radiation of the marine environment (Baxter, 1993).

Burning of coal results mostly in the transfer of radioactivity to the ash, with ^{226}Ra tending to concentrate in the finer fractions (Paridaens and Vanmarcke, 2005). Modern coal-fired power plants use flue gas purification, consisting of a denitrification unit and a desulfurization unit, which complement the electrostatic filter units. At the operating temperature of coal-fired power stations, it is possible for some compounds containing radionuclides to be volatilized. However, the cleaning units ensure that volatile radionuclides are released to the atmosphere only to a small extent, together with a relatively small proportion of the ash. The compounds in gaseous form are free to travel with the flue gases until they cool sufficiently to condense. In this case, particles of ash in the gas stream serve as condensation nuclei for the volatilized nuclides. The result is an enhanced activity concentration of some radionuclides on fly ash (e.g., volatilization of ^{210}Po and ^{210}Pb can cause their condensation on the fly ash (boiling point: 1,740 and 962°C, respectively). Relatively little enhancement by volatilization has been reported for other elements (UNSCEAR, 1988). Generally, enrichment factors increase with decreasing particle size. Examples of mass balance and activity flow in coal-fired power plants may be found in the open literature (Mugrauer et al., 1993; Spears and Martinez-Tarrazona, 2004; Karayigit et al., 2006).

Disequilibrium can be found in coal combustion products like fly ash. As a consequence, fly ash may be enriched in ^{210}Pb and ^{210}Po with respect to the other ^{238}U series radionuclides (Baxter, 1996).

The natural radioactive nuclide concentration in fly ash exceeds that in coal by a factor of 2–15. Only a small percentage of the radioactivity can be found in the FGD products, such as gypsum. In Table 3.17, the activity concentration data of coal fly ash, bituminous boiler slag, and bottom ash are compared.

The highest activity concentrations for ^{226}Ra and ^{232}Th are found in fly ash, followed by boiler slag. The type of coal used has a very significant effect upon the activity discharged into the environment, as does the plant design. Lignite (brown coal, BC) has lower ash content than other coal, which results in a higher enrichment factor. However, lignite has generally lower activity concentration. This is partially compensated by the higher fly ash production from lignite-burning coal power plants.

Table 3.17 Comparison of the activity concentration (mean value) of coal fly ash, bituminous boiler slag, and bottom ash from German coal-fired power plants in Bq kg^{-1}.

Source	^{226}Ra	^{232}Th	^{40}K
		Bq kg^{-1}	
Fly ash	186	110	785
Boiler slag	146	100	794
Bottom ash	108	79	514

Source: Adapted from Puch et al. (2005).

3.2.2.1. Exposure routes

The radiological impact on members of the public can occur due to four main exposure pathways: (1) release to the atmosphere from the stack; (2) release to the atmosphere from ash piles; (3) the use of ash in building materials; and (4) disposal of ash in landfills. As a result of the lack of appropriate legislation, some of these wastes have been dumped in the environment without any regard to radiation protection requirements (Bokoria et al., 2007).

Large amounts of ash are generated by use of coal as a source of energy, for example: Italy about 1 Mt y^{-1} (Trotti et al., 2005); Germany 17.3 Mt y^{-1} (Puch et al., 2005); France about 2 Mt y^{-1} (Degrange and Lepicard, 2004); Greece over 2 Mt y^{-1} (Karangelos et al., 2005).

Table 3.18 shows as an example the production of by-products due to burning bituminous coal and lignite in German coal-fired power plants.

A large amount of the by-products from coal-fired power plants is utilized basically by two industries:

1. *mining*: backfilling of exhausted mines; recultivation of former mining areas; and
2. *construction*: unmixed; mixed with FGD gypsum and/or FDG water.

The main exposure pathways due to the *release to atmosphere* from the stack and piles are: (a) inhalation of radionuclides in the plume; (b) external irradiation by radionuclides in the plume; (c) external irradiation by radionuclides from the plume deposited on the ground; (d) inhalation of radionuclides resuspended from ground deposits; (e) ingestion of food grown on land contaminated by deposited radionuclides; (f) direct external

Table 3.18 Production of boiler slag, bottom ash and fly ash due to burning bituminous coal and lignite in German coal-fired power plants in the year 2000 (Puch et al., 2005).

Type of power station	By-product	Production (Mt)
Bituminous coal power plant	Boiler slag	2.35
	Bottom ash	0.53
	Fly ash	4.12
	Fluidized bed combustion ash	0.42
	Subtotal bituminous coal	*7.42*
Lignite coal power plant	Bottom ash	1.83
	Fly ash	7.90
	Fluidized bed combustion ash	0.19
	Subtotal lignite coal	*9.92*

irradiation from the piles; and (g) exposure from building products using recycled ash.

The primary source term for the radiological assessment is the emission of radionuclides through the stack of a coal-fired power plant. In Table 3.19, the estimated stack emission of different radionuclides is shown for a generic coal-fired power plant.

The largest annual activity emitted into the atmosphere is of ^{222}Rn, followed by ^{210}Pb and ^{210}Po.

Another important component in the radiological assessment is the *fly ash disposal method* used. Fly ash can be disposed of by the following methods (Karangelos et al., 2005):

(a) Fly ash water slurry transportation to an open disposal site using open water cycle systems. This can affect the natural radiation environment of an area in a significant manner, for example, through direct contact with lake water, leaching of radionuclides from fly ash into the lake, with rainwater and pile drainage water acting as the main sources of radioactive contamination. As a result, the ^{226}Ra activity concentration in lake water, in lake sediments and sediments of outflow water are increased. Furthermore, aerial transport of fly ash from the surface of the deposit to the surrounding environment due to wind erosion has to be reckoned with.

(b) Fly ash water slurry transportation to an open disposal site using closed water cycle systems. This represents an improvement over the method described above under (a), but fly ash deposits need to be covered adequately with soil or other material in order to deal with potential pile drainage water or wind erosion. Until 2004, in Hungary about 28 Mt of ash and slag containing elevated levels of naturally occurring materials have been released to the environment as ash and slag mixture in a water suspension into a controlled and monitored area, surrounded by dams (Szerbin et al., 2005). The wastes formed huge ponds, gradually drying as the water from the ponds has been recovered and

Table 3.19 Estimated stack emission of radionuclides into the atmosphere from a reference coal-fired power station (Penfold et al., 1998).

Radionuclide	Stack release ($Bq\,y^{-1}$)
^{238}U series to ^{226}Ra	8.0×10^8 each
^{210}Pb and ^{210}Po	1.6×10^9 each
^{232}Th series	4.0×10^8 each
^{235}U series	4.0×10^7 each
^{222}Rn	1.7×10^{11}

Note: Assumption, 4,000 MW_e plant; effective release height, 300 m.

recycled. The area of the tailing ponds covers approximately $2.7 \times 10^6\ m^2$. The ash and slag mixture was deposited directly on the soil surface without any treatment (thickness: 18–21 m, covered by 30 cm soil).

(c) Disposal of fly ash in sequestered landfills. This approach requires a well constructed and managed site in order to prevent radionuclide migration to underground water and resuspension into air before the site can be offered for recultivation.

(d) Disposal of fly ash in exhausted strip mine fields. This method disposes of fly ash (^{226}Ra sometimes exceeding 1,000 $Bq\,kg^{-2}$) as layers several meters thick, covered by a layer of soil, using end belt transporters and high capacity tracks.

In the long term, exposure of members of the public from radionuclides contained in the wastes deposited can occur either by leaching into groundwater or by intrusion into the landfill (e.g., during construction of foundations). The radiological impact can be increased further by the migration process of the mobile fraction of these radionuclides. Their uptake into the ecosystem can enhance the internal doses to biota. In Poland, saline pit waters resulting from hard coal (HC) mining were originally pumped untreated to the surface and discharged into tailing ponds to allow any suspended material to settle (Leopold and Michalik, 2007). Those waters contained – besides heavy metals – in some cases elevated levels of naturally occurring radionuclides, mainly ^{226}Ra and ^{228}Ra. The radium isotopes were partly deposited as scales inside the discharge tubes or with the sediments of the tailing pond. The scales can contain activity concentrations of more than $1.0 \times 10^5\ Bq\,kg^{-1}$ of ^{226}Ra and more than $6.0 \times 10^4\ Bq\,kg^{-1}$ of ^{228}Ra, whilst the tailings have about $6.0 \times 10^3\ Bq\,kg^{-1}$ of ^{226}Ra and $7.2 \times 10^3\ Bq\,kg^{-1}$ of ^{228}Ra. Since the total volume of such waste material can amount to hundreds of thousands of cubic meters, they can represent a significant radioecological risk.

Table 3.20 shows the activity concentrations in selected deposits, representing mining residues from settling ponds and underground galleries (Chalupnik, 2005). The range of values is significant and can cover three orders of magnitude.

The mine water can be purified, thereby reducing the amount of ^{226}Ra and ^{228}Ra released. In the case of the Polish Piast mine, for example, the rates of release of ^{226}Ra and ^{228}Ra through the Gostynka river, which receives water from the settling pond, into the Vistula river are approximately 50 and 70 $MBq\,d^{-1}$, respectively (Chalupnik, 2005). Purification of radium-bearing waters with regard to the natural environment brings the $^{226}Ra + ^{226}Ra$ concentration one order of magnitude down, from 15 to 1.5 $kBq\,m^{-3}$ (Chalupnik, 2005).

Table 3.20 Activity concentrations in selected deposits due to coal mining in Poland (in $Bq\,kg^{-1} \pm 1\sigma$ error) (Chalupnik, 2005).

	^{226}Ra	^{228}Ra	^{224}Ra	^{40}K
	$Bq\,kg^{-1} \pm 1\sigma$			
Bojszowy Reservoir, near outlet to Gstynka River	344 ± 18	413 ± 18	408 ± 30	530 ± 67
Bojszowy Reservoir, near inflow from zeczott Colliery	$1,844 \pm 110$	$4,391 \pm 177$	735 ± 53	653 ± 89
Bojszowy Reservoir, near inflow from Piast Colliery	442 ± 21	$1,042 \pm 42$	460 ± 45	419 ± 33
Rontok Reservoir, bottom sediment	$8,289 \pm 288$	$3,177 \pm 55$	$2,716 \pm 108$	472 ± 132
Chwalowice Colliery, underground deposits	$35,900 \pm 510$	$25,200 \pm 390$	$20,800 \pm 470$	$1,790 \pm 365$
Piast Colliery, waste rock form the surface disposal pile	83 ± 6	74 ± 3	70 ± 6	724 ± 87

As a consequence of several coal mines having been closed in the recent past, containing thousands of tons of bottom sediments with enhanced natural radioactivity in their settling ponds, radium from such deposits may be leached and may then contaminate the groundwater in the future.

3.2.2.2. Doses

Activity concentration data for brown and hard coals (BC and HC) and their combustion products are summarized in Table 3.21. The uranium and thorium concentrations in BC ranges from 0.5 to 16.3 ppm, and from 0.2 to about 30 ppm, respectively; the corresponding range for uranium in HC is between 2 and 9 ppm. Combustion leads to elevated ^{226}Ra and ^{232}Th levels in ash and slag relative to the coal, whilst gypsum contains only negligible levels of radionuclides, comparable to the Ra content of natural gypsum. Therefore, radon exhalation from a plasterboard manufactured from recycled gypsum is also insignificant: assuming 500 kg of recycled gypsum has been used in a $4 \times 3 \times 2.5\,m^3$ bedroom (which is enough to cover all walls of the room), an exhalation rate of $2 \times 10^{-6}\,Bq\,s^{-1}\,kg^{-1}$ and a ventilation rate of $0.7\,h^{-1}$, this would result under a worst case scenario (all of the emitted radon ends up in room, and 100% residence time in this room) in a radon increase of only $< 0.2\,Bq\,m^{-3}$ and a thoron EEC of

Table 3.21 Activity concentration (in Bq kg^{-1}) of Czech brown coal (BC) and hard coal (HC) used in coal-fired power plants and of the combustion products.

Radionuclide	BC–ash	BC–ash	BC–gypsum	HC–ash	HC–slag	HC–gypsum
^{226}Ra	84–139	67–81	<9	129	78	<6
^{232}Th	77–92	50–64	<9	82	59	<9
^{40}K	169–355	<28–289	<26	700	637	<13

Source: Adapted from Moravanska and Laciok (2004).

0.01 Bq m^{-3}. The corresponding annual effective dose equals 3 μSv (Paridaens and Vanmarcke, 2005).

A large proportion of the ash generated in a coal-fired power plant is sold – for example, in Germany 99.9% of the 17.34 Mt of by-products are utilized (Puch et al., 2005) – and used as building materials, cement or light-weight filler for concrete. Therefore, these products represent another potential source term to be considered. Using bottom or fly ash as additives in concrete, it is necessary to assess the effect on radon exhalation (Paridaens and Vanmarcke, 2005). Current practice foresees a maximum addition of 18% (volume) of fly ash to cement. Cement is present at 30% (volume) in concrete – i.e., this represents a 3% (weight) fraction of fly ash in concrete. Normal concrete without fly ash has an activity concentration of about 30 Bq kg^{-1} of ^{226}Ra. Adding 3% weight fraction of fly ash containing 200 Bq kg^{-1} of ^{226}Ra would increase the ^{226}Ra activity concentration in the concrete by about 20%; increasing ^{226}Ra activity concentration through adding fly ash by 80% would lead to an increase of the radon exhalation by 20%. Since the contribution to radon indoors by concrete in a room with *all* walls being concrete is estimated to be a few tens of Bq m^{-3}, an increase of 18% would only amount to a few Bq m^{-3} that can be considered as an insignificant dose contribution to the dose from NRE.

In the following section, the *non-occupational radiological impact* of a generic coal-fired power plant on members of the public is summarized (Penfold et al., 1998). The following assumptions have been made for the generic coal-fired power plant:

1. *Energy production*: 2.8×10^4 GWh y^{-1} of electric power, consuming 1×10^7 t y^{-1} of coal and generating 1.7×10^6 t y^{-1} of ash. This would result in a release of 8×10^3 t y^{-1} of ash into the atmosphere – i.e., 0.5% of the total ash produced (the efficiency of electrostatic filters is high, and typically >99.5%).
2. *Atmospheric releases:* plume released at 300 m effective release height from a 200 m high stack due to initial buoyancy and momentum of the hot flue gases.

3. *Radionuclide concentration in fly ash:* ^{238}U series to ^{226}Ra 100 Bq kg^{-1} (each), ^{210}Pb and ^{210}Po 200 Bq kg^{-1} (each, i.e., enrichment factor 2), ^{232}Th series 50 Bq kg^{-1} (each), and ^{235}U series 5 Bq kg^{-1} (each).
4. *Ash pile:* 1×10^5 m^3 of ash on site, resulting in a pile 10 m deep, with a surface area of 1×10^4 m^2 (atmospheric concentration of resuspended ash above the pile: 1×10^{-7} kg m^{-3}; radon emanation fraction: 5.4×10^{-3}; radon diffusion coefficient: 5×10^{-6} m^2 s^{-1}); build-up and attenuation characteristics of ash similar to concrete (density 0.75 t m^{-3}).
5. *Ash disposal in landfill*: basic landfill with no lining and limited capping (0.5 m of soil); capacity 5×10^6 m^3, covering an area of 100 ha (average depth: 5 m).
6. *Exposure from building products*: building material is 25% ash by mass; exposure is due to external irradiation from radionuclides contained in the structure of the house and due to inhalation of radon emanating from its walls; ventilation rate is 0.5 h^{-1}.
7. *Critical group:* 25% of food consumed from local area, grown on land contaminated by deposited radionuclides, eating the two most significant foodstuffs at critical rates and all others at average rates; direct irradiation from ash piles; inhalation of radionuclides from the plume; adult lives 500 m from the stack or ash piles; inside residence time 50% of the year at that location.

In Table 3.22, the effective doses for a member of the critical group from releases of the generic coal-fired power plant are summarized (Penfold et al., 1998). The highest dose 250 μSv y^{-1} is received due to the use of fly ash in constructing dwellings. Stack releases and releases from ash piles

Table 3.22 Non-occupational exposure by a member of the critical group due to releases of a generic coal-fired power plant.

Exposure scenario	Total effective dose (μSv y^{-1})	Dominant exposure pathways and radionuclides	Contribution of total effective dose (%)
Stack releases	55	^{210}Pb and ^{210}Po	99
Ash piles	6.9×10^{-2}	^{210}Pb and ^{210}Po	49
		^{238}U series and ^{232}Th series direct external irradiation	34
Landfill disposal	16	^{238}U series and ^{232}Th series	44, resp. 56
		Radionuclide migration (peak after approximately 9×10^4 y)	
Building materials	250	^{222}Rn and ^{220}Rn inhalation	96
		External irradiation	4

Source: Adapted from Penfold et al. (1998).

impact individuals mainly through the contamination of foodstuffs (green vegetables and grain); in the case of ash piles, external irradiation is an additional dose component. Landfills contribute to the dose only in the long term, that is, due to leaching of radionuclides from the ash into groundwater. Since chemically the fly ash is a silicate–aluminate with low calcium content, such transport processes will only occur after about 9.0×10^4 y.

Occupational exposure of workers in a coal-fired power plant occurs mostly through fly ash. Usually fly ash is sent from the power station to the end-user in the construction industry immediately after being produced. However, due to the seasonality of some construction work (e.g., relatively few road works during the winter months), fly ash may be temporarily stored in basins or on tips. Below the radiation exposure at a generic coal-fired power plant is assessed for the following occupational exposure scenarios (Degrange and Lepicard, 2004):

1. a worker who operates periodically in the boiler ash box, which is used for retrieving bottom ash and is filled with an ash water slurry;
2. an ash handling agent who spends time during fly ash loading onto a truck or during fly ash humidification before loading;
3. a person who spends time at storage silos for dry fly ash beside hoppers under electrostatic precipitators, conducting surveillance rounds, or during repair or maintenance operations;
4. a worker who repairs an electrostatic precipitator;
5. a worker who operates a vehicle on the fly ash tip for continuous fly ash levelling or truck loading;
6. a worker who operates a truck, loading or unloading dry or wet fly ash;
7. a worker who is exposed to fly ash storage silos and loaded concrete trucks in a concrete factory;
8. a worker in a building material factory who operates close to a pile of construction material containing fly ash as partial substitute for cement; and
9. a road worker who is exposed to fly ash used as embankments or as aggregates in road sublayers.

The exposure conditions and the source-term fly ash are assumed to have the following characteristics: (a) air dust load – $10\,mg\,m^{-3}$; (b) radionuclide-mean activity concentration ($Bq\,kg^{-1}$): ^{226}Ra – 138, ^{232}Th – 119.3, ^{210}Po and ^{210}Pb – 116.8 (each). Table 3.23 contains the annual doses for the above occupational exposure scenarios.

The main contribution to the external radiation exposure results from members of the ^{232}Th decay chain segments ((^{228}Ra, ^{228}Ac) and (^{228}Th to ^{208}Tl) represent approximately half of the total external exposure). Members of the ^{238}U decay chain (segment (^{226}Ra to ^{214}Po)) deliver about 30% of the external exposure. K-40 represents 23% of the exposure.

Table 3.23 Annual effective doses for various occupational exposure scenarios in a generic coal-fired power plant described in the text above.

Exposure scenario no.	Dose ($\mu Sv\,y^{-1}$)	Comments
1–4, 6–8	<500	
5	500	60% from external irradiation
9	310	54% from external irradiation

Source: Adapted from Degrange and Lepicard (2004).

Th-228, ^{232}Th, and ^{224}Ra as members of the ^{232}Th decay chain contribute about 60% of the total exposure from airborne dust inhalation; ^{40}K does not contribute significantly to this exposure route.

Radon inhalation contributes at most 25% of the total exposure; in this case, the exposure needs to be indoors, as is the case for an employee in the assumed building material factory.

3.2.3. Mineral sands

3.2.3.1. Brief history

Well before the phenomenon known today as radioactivity was discovered, a liquid mixture of thorium, lanthanum, and cerium was used to embed small fabric bags used to enhance the brightness of street lamps in the largest cities of Europe. This mixture was obtained from monazite sands coming from several parts of the world, but mostly from Brazil and India.

At the beginning of the 20th century, thus just after the discovery of radioactivity, thorium started being separated from monazite sands for various purposes, including for therapeutic applications, like thorotrast – a stabilized colloidal thorium dioxide used as contrast for roentgenography. The remnants of those aforementioned operations would be called today monazite cycle derived NORM or TENORM.

The heavy mineral sands are the source of zirconium, titanium, thorium, tungsten, and rare earth elements. Yttrium ($Z=39$), an element that follows the lanthanides in the table of elements, is usually also present in these sands, as well as scandium ($Z=39$). Here it is worth remembering that rare earths mean a group of 15 elements (i.e., $57 \leq Z \leq 71$). Most, if not all, rare earths have commercially important applications. As a consequence, minerals like monazite, bastnaesite, cerites, xenotime, gadolinite, among others are extracted in industrial operations.

Table 3.24 lists the composition of selected heavy minerals, which are extracted for industrial applications.

In general, monazites are rare earth phosphates and are found throughout the world. The origin, age, and composition of monazites

Table 3.24 Typical percentage compositions of selected heavy minerals from several origins (Paschoa, 1994; Pillai, 2008).

Mineral	Origin	Rare earth oxides	U_3O_8	ThO_2	P_2O_5
Monazite	Australia	61.3	0.34	6.6	26.3
	Brazil	59.2	0.17	6.5	26.0
	India	59.7	0.37	9.6	26.2
	Malaysia	59.7	0.24	5.9	25.7
	Thailand	60.2	0.44	5.8	26.5
	USA	40.7	0.47	3.1	19.3
Bastnaesite	Several	58–74	–	0.11–0.20	0.64–0.94
Xenotime	Malaysia	54.0	0.81	0.83	26.2
Gadolinite	Several	32–46	–	≤ 2	–

are related. Pleistocene and Quaternary sediments are from distinct origins (Machado and Gauthier, 1996; Yang et al., 2006; Dawood and Abd El-Naby, 2007). Rain precipitation associated with floods and the variations of the sea coast are usually considered to be responsible for the erosion of monazite source rocks. Bastnaesite is a cerium lanthanum yttrium carbonate fluoride – Ce, La, Y (CO_3F), which like monazites have related origin, age, and composition (Olson et al., 1954; Hughe and Brown, 1977). Gadolinite is a rare mineral, which appears as attractive crystals in the chemical form of yttrium iron beryllium silicate ($Y_2FeBe_2Si_2O_{10}$).

The beginning of the WWII saw an increase in applications of monazite from kerosene lamps using fabric bags impregnated with thorium nitrate to small flints used in wind-resistant lighters widely adopted by army soldiers, mariners, and air-force personnel to light their cigarettes. At the end of WWII and in subsequent years, monazite had become a strategic commodity not only because of the extraction of rare earth compounds used in various industries but also due to the possibility of using thorium in the planned breeder nuclear reactor program. Late in the 1960s, the breeder lost momentum and monazite was extracted and processed mostly to obtain its rare earth and other compounds widely used in sophisticated products like the then new color TV, modern airplanes, and spacecraft (see e.g. Paschoa, 1994).

The main components of the cerium group (cerium, lanthanum, neodymium, praseodymium, and samarium), but also europium and thulium, were at first considered contaminants, usually in the form of chlorides, in the earlier processes to extract thorium and uranium from monazite. The selling price of mixed rare earths increased from \$1.5 kg^{-1} in 1964 to \$2,000 kg^{-1} in 1987 (Padamanabhan, 1987).

Most methods or combination of procedures used commercially for the separation of thorium, uranium, and rare earths may be described summarily as follows. After opening up the monazite mineral for treatment with sodium hydroxide at high temperature, it is digested in sulfuric acid (Kremers, 1949; Paschoa, 1994). Figures 3.5 and 3.6 illustrate the typical two phases of the monazite cycle used in several industries. Figure 3.5 represents a simplification of phase I (physical separation) of a typical monazite cycle, which consists of extraction, concentration, electrostatic, and magnetic separations. Phase I includes extracting monazite-bearing sands from the beaches, washing them, and concentrating them by hydrogravimetry (using Humphrey's spirals), drying, and then the physical separation processes. The products after phase I are: magnetic ilmenite; nonmagnetic rutile; magnetic monazite; and nonmagnetic zirconite. Figure 3.6 represents phase II (chemical treatment) of the monazite cycle. It is in phase II that the NORM wastes from the monazite cycle appear in the form of cake II (Th and U oxides), and mesothorium (Ba, Ra SO_4). After a few steps that involve further magnetic separation and removal of ilmenite, ilmenite plus monazite, and zirconite fractions, the concentration of monazite reaches about 99%. This highly concentrated monazite is ground and subsequently submitted to an alkaline attack and filtration. Sodium phosphate and a mixture, usually called cake I, are obtained at this stage. While sodium phosphate is crystallized and centrifuged to yield tri-sodium phosphate and sodium hydroxide (caustic soda), cake I is homogenized, neutralized, and filtered. This latter treatment of cake I yields a filtrate containing rare earth chlorides and a new mixture called cake II. Most of the initial thorium and uranium contents of monazite concentrates remain

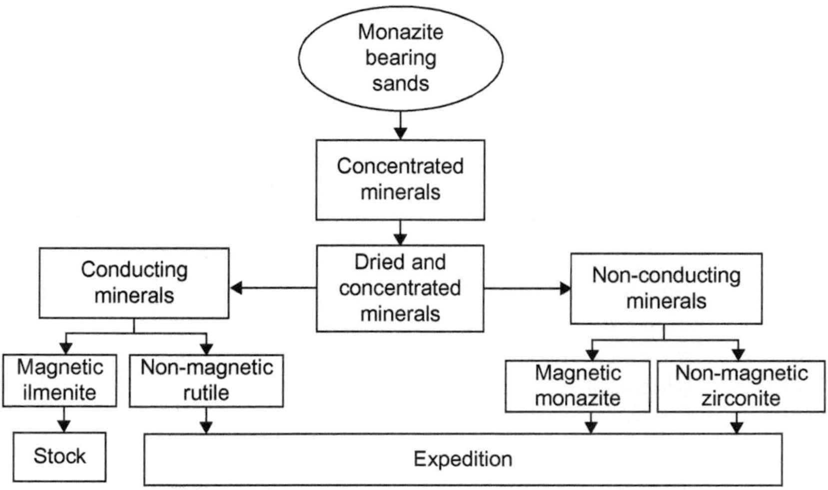

Figure 3.5 Schematic representation of phase I (physical separation) of a typical monazite cycle. Adapted from Paschoa (1994).

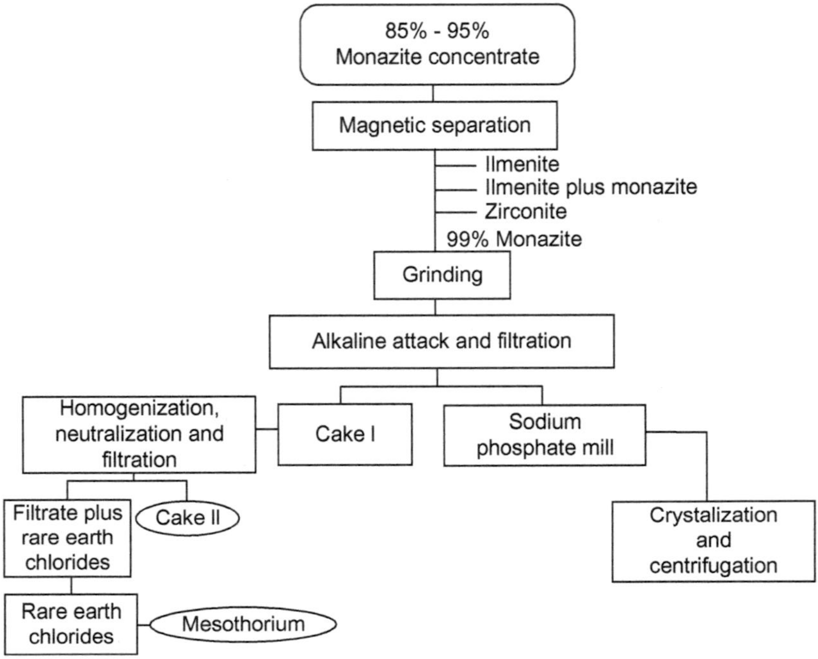

Figure 3.6 Schematic representation of phase II (chemical treatment) of a typical monazite cycle. Adapted from Paschoa (1994).

in cake II. As a consequence, a TENORM problem occurs in the monazite cycle because large volumes of cake II are produced.

The rare earth chlorides are obtained by means of the following reactions (Pillai, 2008):

$$RE(PO_4) + 3NaOH \rightarrow RE(OH)_3 + Na_3PO_4$$

$$Th_3(PO_4)_4 + 12NaOH \rightarrow 3Th(OH)_4 + 4Na_3PO_4$$

$$RE(OH)_3 + 3HCl \rightarrow RECl_3 + 3H_2O$$

Radium isotopes (mostly ^{226}Ra and ^{228}Ra) present in rare earths are carried by barium sulfate and coprecipitated with barium chloride with sodium sulfate (Pillai, 2008).

For each 100 mt of monazite chemically treated, around 2% of cake II, and 6–7% of mesothorium are produced (Paschoa, 1993). Cake II and mesothorium (mostly ^{228}Ra compounds) are grossly separated from each other by means of precipitating thorium and subsequent filtration.

In Brazil alone, the volume of monazite-related TENORM wastes had reached a volume of 1×10^4 m^3 until the year 2001, after about 60 years of operations. The wastes are kept in drums, storage buildings, concrete silos, and trenches. In addition to Brazil, it is known that there are historic NORM-bearing wastes from mineral sands operations in Australia, China, Egypt, India, Malaysia, South Africa, Sri Lanka, Taiwan, and the United States.

3.2.3.2. NORM in the mineral sands industry

The monazite cycle and other mineral sands industries leave behind residues containing NORM with combined ^{226}Ra plus ^{228}Ra activity concentrations which reach levels as high as 10^4 kBq kg^{-1}.

These NORM wastes from the monazite cycle result from grinding and chemical processes, which make these radium isotopes more bioavailable than those from the non-chemical processes (Paschoa, 2008). The latter is the case, for example, for the NORM wastes that appear in the oil and gas industry. A discussion of the long-term behavior of ^{226}Ra plus ^{228}Ra in the remnants of the monazite cycle is presented in Section 3.1.5.

Table 3.25 presents the ranges of ^{228}Ra in solid and liquid wastes from the monazite cycle, and the external radiation fields, as reported for the monazite cycle in India (Pillai, 2008). Activity concentrations as high as 5.0×10^4 Bq kg^{-1} were determined in mixed cake of the monazite cycle. For each metric ton of monazite processed, about 2.5×10^2 kg of thorium hydroxide cake and 15 m^3 of liquid effluents are produced. Considering a plant processing between 3.0×10^3 and 4.0×10^3 mt y^{-1}, the total amount of thorium-bearing cake would be from 7.5×10^2 to 1.0×10^3 mt, and the volume of liquid effluents would be between 4.5×10^4 and 6.0×10^4 m^3. It was estimated that the individual worker dose over the past 25 years was in the range 3–9 mSv y^{-1} (Paschoa, 2008).

Table 3.25 Ranges of ^{228}Ra activity concentrations in solid wastes and liquid effluents from the monazite cycle, and the associated external radiation fields, as reported for the monazite cycle in India.

Solid wastes	^{228}Ra (Bq kg^{-1})	Radiation field (μGy h^{-1})
Monazite, insoluble	4.0×10^5–1.0×10^6	30–1.0×10^2
Mixed cake	2.0×10^6–5.0×10^6	4.0×10^2–6.0×10^2
Effluent treatment plant effluent	2.5×10^4–1.0×10^5	2–3
Liquid effluents	^{228}Ra (Bq m^{-3})	
Acid effluent (pH 1.6–2.0)	1.5×10^5–2.0×10^5	–
Alkaline effluent (pH 12–13)	3.0×10^5–4.0×10^5	–

Source: Adapted from Pillai (2008).

Titanium oxide (TiO_2) is a white powder with a very high refractive index (2.7 average), which implies high opacity, brilliant whiteness, and covering power. There are two types of titanium oxides: (i) TiO_2, rutile, refractive index 2.80 and (ii) TiO_2, anatase, refractive index 2.55 (Buchel et al., 2000). Titanium oxides are obtained from minerals containing rutile or ilmenite. Rutile is an impure form of TiO_2, whereas ilmenite is a compound oxide of Fe and Ti. The minerals from which titanium oxides are extracted contain ^{232}Th, ^{228}Ra, plus measurable amounts of ^{226}Ra. The annual external gamma exposure rates and the doses from airborne thorium levels are typically of the order of 1 and 0.7 mSv (Haridasan et al., 2008). Information on the long-term radiological impact of ^{232}Th, ^{228}Ra, and ^{226}Ra based on direct measurements on wastes produced in the titanium dioxide pigment industry is scarce.

There are two types of processes for producing TiO_2: the chloride process; and the sulfate process. Figures 3.7 and 3.8 present the flow charts of the chloride and the sulfate processes, respectively.

The chlorine process is represented by the following chemical reactions (McNulty, 2008):

$$TiO_2 + 2Cl_2 \rightarrow TiCl_4 + CO + CO_2$$

$$TiCl_4 + O_2 \rightarrow TiO_2 + 2Cl_2$$

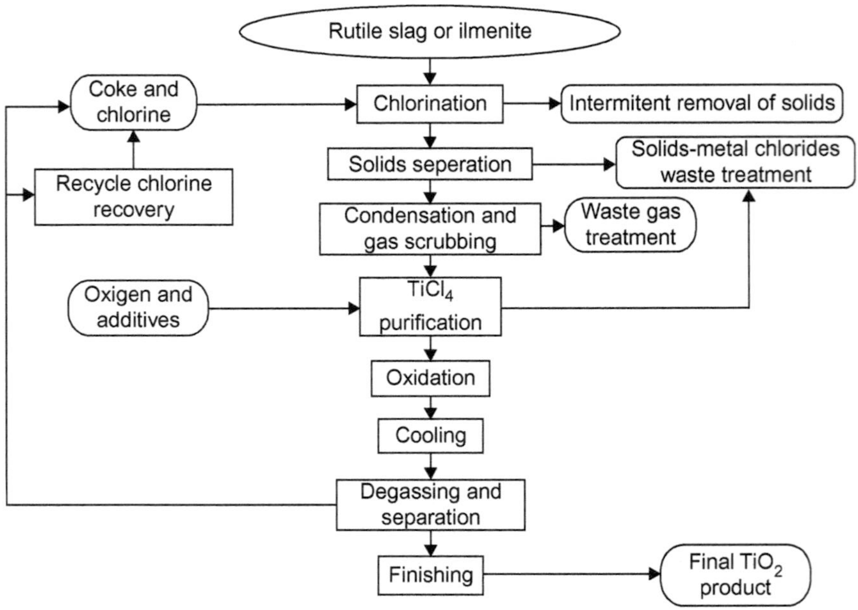

Figure 3.7 Flow chart of the chlorine process to produce TiO_2. Adapted from McNulty (2008).

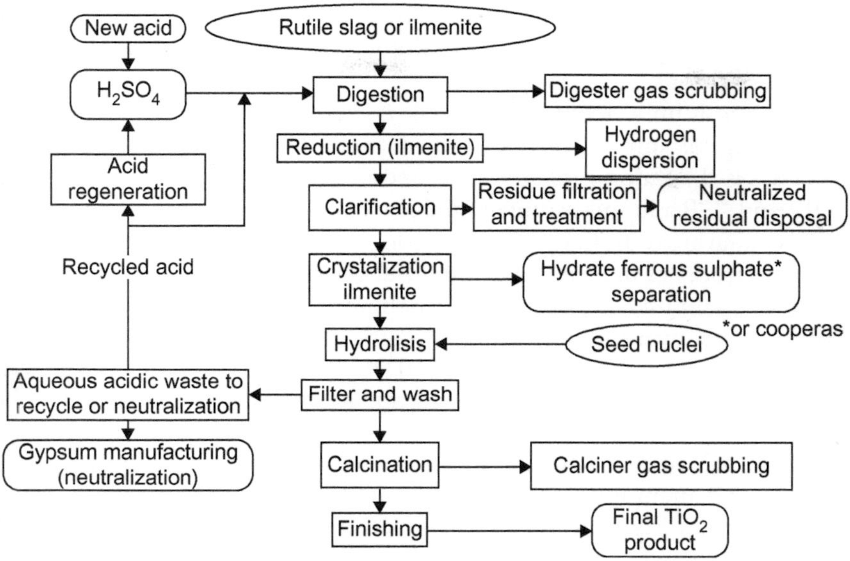

Figure 3.8 Flow chart of the sulfate process to produce TiO_2. Adapted from McNulty (2008).

The NORM wastes in the chlorine process are produced in the intermittent removal of solids, in the gas treatment and in the solids-metal chlorides treatment. An amount of radon gas is released during the chlorine process and is dispersed through the surrounding environment. This almost continuous radon release may enhance the ^{210}Pb (22.3 years half-life) and ^{210}Po (138.4 days half-life) in the surrounding environment, with possible accumulation in some species of flora and fauna. The radiological impact of the chlorine process has been recently discussed (McNulty, 2008). However, it is advisable to investigate such impacts based on site-specific situations.

The sulfate process is represented by the following chemical reactions (McNulty, 2008):

$$FeTiO_2 + 2H_2SO_4 \rightarrow TiOSO_4 + FeSO_4 + H_2O$$

$$TiOSO_4 + H_2O \rightarrow TiO_2 + nH_2O + H_2SO_4$$

$$TiO_2 + nH_2O \rightarrow TiO_2 + nH_2O$$

The NORM wastes in the sulfate process are produced in the crystallization, hydrolysis, and filtration and washing phases. The aqueous acidic waste and gypsum receive most of the NORM produced in the sulfate process. The radiological impact of the sulfate process has been recently discussed (McNulty, 2008). However, as in the case of the chlorine process, it is advisable to investigate the radiological impact based on site-specific situations.

The most important aspect as far as the long-term radiological and radioecological impacts are concerned in both the chlorine and sulfate processes is the radon progeny deposited in the surrounding environment, ^{228}Ra and ^{226}Ra in soils, aquifers, and seawater, if the operations occur near or on coastal areas. In the past, it was common to dispose of waste from the titanium processes in nearby sea water but this practice has been abandoned by now in almost all cases, leaving behind a radioecological impact that remains to be investigated. Here, it is worth remembering that plankton has concentration factors of the order of 10^4 to 10^5 (Paschoa et al., 1981). The aquatic food chain may have received impacts of radium releases yet to be understood.

Table 3.26 lists the typical activity concentration values and ranges of thorium and uranium (without further information) in ilmenite and two kinds of wastes from the titanium pigment industry in Australia, and their respective exempt activity concentrations. The ^{226}Ra activity concentration in ilmenite from Cox's Bazar area in Bangladesh is 3.5×10^2 Bq kg^{-1} (Alan et al., 1999), which lies within the range shown for ^{238}U activity concentration in Table 3.26.

According to Table 3.26, the values and ranges found in ilmenite and waste materials in the pigment industry in Australia are slightly below or slightly above the exemption activity concentration of 1×10^3 Bq kg^{-1}. Data on ^{226}Ra activity concentrations in wastes of the titanium industry elsewhere have not been found; however, it has been reported that levels as

Table 3.26 Typical uranium and thorium activity concentration values and ranges in ilmenite and wastes of the Australian titanium pigment industry are of the same order of magnitude, while some are higher than their respective exempt concentrations.

	Activity concentration	Exempt activity concentration
	Bq kg^{-1}	
^{232}Th	4.0×10^2–4.0×10^3	1×10^3
^{238}U	$<2.5 \times 10^2$–8.0×10^3	1×10^3
^{232}Th	2.5×10^3	1×10^3
^{238}U	7.5×10^3	1×10^3
^{232}Th	1.9×10^3–2.9×10^3	1×10^3
^{238}U	7.5×10^2–1×10^3	1×10^3

Source: Adapted from Koperski and Tsurikov (1999).

high as $2.9 \times 10^3\,Bq\,kg^{-1}$ can occur (i.e., the same as that of ^{238}U in filter cake shown in Table 3.26).

After the Council Directive 96/29/EURATOM of 13 May (EURATOM, 1996) was integrated into the national legislation of some European countries, some of these countries decided to adopt a three-phase plan for the NORM industry to adapt and implement the legislation into the processes, operations, and waste treatment and disposal. The first (or initial) phase was just to understand the implications of the Council Directive 96/29, and how it was applied in other countries. The second (or intermediate) phase was used to design regulatory criteria, and analyzed their implications for the country's NORM industries. The third (or implementation) phase, usually not yet achieved in most countries, will be the implementation of the legislation into the processes, operations, and wastes of the NORM industry. The development of these three phases may be further complicated by the fact that the residues of NORM industries sometimes might have been disposed of at conventional wastes repositories or might have been reused in uncontrolled manner in a myriad of applications. Whenever possible it may be advisable to change industrial processes and operational procedures, and introduce radiological control measures and monitoring into existing conventional repositories.

One added problem is that world demand for titanium dioxide and titanium alloys is increasing rather than declining. Thus the amount of wastes from the titanium industry is tending to increase annually, unless a safe reuse of such wastes can be identified in the foreseeable future.

3.2.3.3. Gas mantle industry

As remarked in Section 3.2.3.1, a liquid mixture of thorium, lanthanum, and cerium was used since a long time back to embed small fabric bags used to enhance the brightness of street lamps. These thoriated gas mantles are still in use today. The radiologically relevant thorium isotopes ^{232}Th, ^{230}Th, and ^{228}Th are present in the mixture used to embed gas mantles. An investigation carried out by the German Federal Office for Radiation Protection has estimated that, based on the maximum thorium excretion rate, an annual uptake of about $2.0 \times 10^3\,Bq$ might have occurred in a worker (Ludwig et al., 2008). The external dose rates ranged from near-zero up to $6.8\,mSv\,y^{-1}$ (Ludwig et al., 2008).

Although the doses received by the public at large and by most of the workers in the gas mantle industry are not huge, industrial hygiene in this industry needs to be kept at the highest level to avoid high dose rates. Inhalation of thoron on a daily basis may be considered in cases in which industrial hygiene is not maintained at high standards. In the past, a gas mantle fabrication process in Brazil, which used thorium nitrate in a tank inside a room with poor air circulation, the workers were exposed to a thoron (^{220}Rn) activity concentration of $2.22 \times 10^5\,Bq\,m^{-3}$. As a consequence of 8 hours of inhalation of ^{220}Rn per day, these workers had everyday ^{212}Pb body burden peaks of about 1.1 kBq (Paschoa, 1990). For comparison purposes, the typical ^{220}Rn activity concentration in Guarapari, one of the Brazilian areas of high natural radioactivity, is $2.4 \times 10^2\,Bq\,m^{-3}$ (Paschoa and Pohl-Rülling, 2005).

3.2.4. Phosphate fertilizer industry

One of the consequences of the European Union Council Directive 96/29/EURATOM issued in 1996 (EURATOM, 1996) was a concerted international effort to estimate, and to the extent possible, prevent the potential unwanted impact of the phosphate fertilizer industry. The "Institut Mondial du Phosphate – IMPHOS" (i.e., World Institute of Phosphate) was created to review the impact of the Directive 96/29 on the phosphate industry.

It is known that phosphogypsum (PG) (mostly $CaSO_4 2H_2O$) from several parts of the world contains heavy metals and NORM associated with the original phosphate rocks (see, for example: van der Heijde et al., 1988; Luther et al., 1993; Manyama Makweba and Holm, 1993; García-León et al., 1995; Hull and Burnett, 1996; Rutherford et al., 1994, 1995a, 1995b, 1996; Silva et al., 2002; Shukla et al., 2005; El Afifi et al., 2009).

The activity concentrations of phosphate rocks vary considerably from country to country and even within the same country due to local geology. Table 3.27 lists the uranium and thorium activity concentrations in phosphate rocks used in the phosphate industry of Australia. The highest uranium activity concentrations are in the rocks from Florida and Jordan, whereas the rocks from Australia have uranium concentrations one order of magnitude lower than those from other countries.

Table 3.27 Range of activity concentrations of uranium and thorium used in the Australian phosphate industry.

	Uranium	Thorium
	$Bq\,kg^{-1}$	
Australia	$15–9.0 \times 10^2$	5–47
Florida (USA)	$1.5 \times 10^3–1.9 \times 10^3$	16–59
Morocco	$1.5 \times 10^3–1.7 \times 10^3$	$10–2.0 \times 10^2$
Jordan	$1.3 \times 10^3–1.9 \times 10^3$	–

Source: Adapted from Cooper (2005).

A PG fact-finding forum was held in Florida in 1996 (http://fipr1.state.fl.us/FIPR/FIPR1.nsf/a1380a2dc3df745f85256b4b006398eb/f87c45fa330065da85256b2e00657e5d/$FILE/01-132-117Final.pdf). That forum essentially concluded the following:

1. There is no significant risk to human health for those individuals working with or around PG; for example, those building a road or spreading it agriculturally.
2. Whereas shielding materials placed over PG are able to protect people rather completely from the impacts of radon, they are less effective in attenuating gamma radiation, so that the gamma factor must always be considered.
3. There appears to be no significant risk to someone traveling over a PG-based road, but only to someone who might build a house on it and occupy that house for 70 years (EPA scenario).
4. The marine studies at Louisiana State University employing PG/cement structures as oyster clutch substrates or artificial reefs have shown no bioaccumulation of toxic materials to date, but these studies need to be extended over longer periods to give confidence in the results.
5. The use of PG as a cover material in municipal waste landfills accelerates the decomposition of "standard garbage" and has the promise of providing additional landfill space, but it also increases the evolution of hydrogen sulfide. There is technology available to recover economically the sulfur values of the hydrogen sulfide.

Since 1996 a large number of studies and investigations on PG have been undertaken throughout the world. The EMRAS NORM Working Group final report is just one of them. One interesting proposal, among many, was to take advantage of the current knowledge on uranium mill tailings, which were intensively studied in recent decades to assist in assessment of the following (Landa, 2004): (i) mineral hosts of natural radionuclides; (ii) use of soil analogues to predict the long-term fate of natural radionuclides; (iii) microbial and diagenetic processes that may alter the mobility of natural radionuclides in the surficial environment; (iv) waste management practices and technologies to limit natural radionuclide migration; and (v) the impact of natural radionuclides from the uranium mill tailings on biota. Landa's suggestion may be applicable to many kinds of NORM/TENORM/TENR, including PG.

One can see from Table 3.28 that most radionuclides listed have activity concentrations in the orders of $10\text{–}10^2\,\text{Bq}\,\text{kg}^{-1}$, with ^{226}Ra having the highest concentrations. Here, it is worth mentioning that there is an ongoing debate as to whether or not the large volume of PG generated in the phosphate industry may or may not be put to positive uses such as: as a conditioner for clay soils or soils with high levels of sodium; as a nutrient for growing a wide variety of plants, together with NPK (sodium, phosphorous, and potassium) fertilizers; as an additive in the production of cement; to produce sulfuric acid; as artificial reefs and oyster subtracts; for building block

Table 3.28 The ranges of activity concentrations in phosphogypsum (PG) produced in Brazil (Silva et al., 2002), and in Australia (Cooper, 2005).

	PG – Brazil	PG – Australia
	$Bq\,kg^{-1}$	
^{238}U	1.1×10^2–1.3×10^2	10–24
^{226}Ra	4.0×10^2–8.9×10^2	2.8×10^2–3.5×10^2
^{210}Pb	1.7×10^2–4.9×10^2	3.2×10^2–4.4×10^2
^{210}Po	Not available	1.5×10^2–3.6×10^2
^{232}Th	1.7×10^2–1.9×10^2	4–7
^{228}Ra	2.1×10^2–2.4×10^2	Not available

fabrication; these among others. The phosphate industry has been looking carefully into the advantages and disadvantages of each of these potential uses of PG for many years, because of the enormous amounts being produced each year. If, instead of being considered as waste, PG would be useful for a practical purpose, the phosphate industry would experience a significant economic improvement.

3.2.5. Recycling industry

Scrap metals from the oil and gas industry and almost all other industries mentioned in Section 3.2 are often remelted for further use. This is a procedure that favors the creation of contaminated metals, which are then incorporated in consumer products or in machinery and equipment for other industries. Until some time ago, the risk to the public resulting from remelting NORM-contaminated metals was considered to be minimal. Today, however, it is known that ^{226}Ra in scrap metals is usually responsible for more than half of the detected radioactive contamination of scrap metal. The NORM II International Symposium held in Krefeld, Germany, November 10–13, 1998, which was a follow-up meeting to the Amsterdam NORM Conference, September 8–10, 1997, concluded that the steel industry had to become more aware of the NORM problem in melting operations. Although NORM in scrap metal *per se* was not considered a major problem, more and more NORM-contaminated steel had been refused several times, irrespective of the activity concentrations in it. The Krefeld Symposium was supported by Siempelkamp Nuklear- und Umwelttechnik GmbH & Co, a local scrap metal melting industry with a great deal of experience in the field. In the WM'99 Conference, investigators from Siempelkamp presented a report on the problem of melting NORM-contaminated equipment of an offshore oil platform (Sappok et al., 1999).

Table 3.29 presents the ranges of activity concentrations of selected naturally occurring radionuclides in dust and slag of a melting campaign. According to Table 3.29, the highest ^{210}Pb activity concentration in dust is 1.0×10^5 Bq kg^{-1}, which is a value lower than the German limit of

Table 3.29 Activity concentration ranges and averages (within parenthesis) of selected naturally occurring radionuclides in dust and slag.

	Dust		Slag	
	Bq kg^{-1}			
^{238}U series	Sappok et al. (1999)	Trotti et al. (2008)	Sappok et al. (1999)	Tsai et al. (2003)
^{238}U	–	8–27	–	–
^{226}Ra	3.6×10^2–8.4×10^2 (5.1×10^2)	13–32	1.3×10^3–1.9×10^4 (7.6×10^3)	–
^{214}Pb	1.3×10^2–2.8×10^2 (2.1×10^2)	–	6.0×10^2–9.7×10^3 (4.1×10^3)	5.6×10^3–9.2×10^3 (7.1×10^2)
^{214}Bi	1.1×10^2–2.3×10^2 (2.2×10^2)	–	6.0×10^2–9.0×10^3 (3.8×10^2)	3.8×10^3–6.6×10^3 (5.1×10^3)
^{210}Pb	1.8×10^4–1.0×10^5 (4.8×10^4)	14–4.7×10^4[a]	–	–
^{210}Po	–	2–4.3×10^4[a]	–	–
^{232}Th series				
^{232}Th	–	5–11	–	–
^{228}Ac	–	–	–	1.4×10^2–5.7×10^2 (3.6×10^2)
^{224}Ra	1.0×10^2–1.4×10^2 (1.2×10^2)	–	4.1×10^3–1.1×10^4 (7.4×10^3)	–
^{212}Pb	40–1.1×10^2 (72)	–	3.3×10^2–1.1×10^4 (4.2×10^3)	2.9×10^2–7.2×10^2 (4.9×10^2)
^{212}Bi	70–2.2×10^2 (1.3×10^2)	–	6.2×10^2–1.1×10^4 (4.7×10^3)	84–6.9×10^2 (4.4×10^2)
^{208}Tl	60–1.1×10^2 (80)	–	4.5×10^2–1.0×10^4 (4.0×10^3)	1.1×10^2–2.4×10^2 (1.6×10^2)
^{40}K	–	30–6.2×10^3[a]	–	–

Source: Data adapted from Sappok et al. (1999), Germany, Tsai et al. (2003), Taiwan, and Trotti et al. (2008).

[a]Weep electrostatic precipitator.

5.0×10^5 Bq kg^{-1} (see Sappok et al., 1999). The high ^{210}Pb activity concentration in dust may indicate that occupational monitoring may be required to make a decision on whether this is a significant occupational value or not. In any case, how much airborne dust containing ^{210}Pb may be inhaled is a matter for an interesting investigation. As far as NORM-contaminated slag is concerned, there is a report which states that recycling and disposal of such slag will lead to doses to members of the general public well below the 1 mSv y^{-1} German criterion (Quade et al., 2003). By the same token, one can consider also that the NORM-contaminated slag from Taiwan is below radiological concern.

In accordance with the data presented in Table 3.29, recycling scrap metal from structures and equipment from decommissioning platforms of the oil and gas industry does not appear to be a problem of major concern from the viewpoint of radiological protection, once proper measures are taken at the recycling plant. However, one must bear in mind that approaches and limits vary from country to country, making the transboundary transport of recycled metal difficult.

One way to deal with the problem of NORM-contaminated scrap metal and its derived products is to establish well-understood international regulations and to actively enforce them throughout the world with fines for violators. It will be necessary to define clearly the exemption levels in accordance with already accepted NORM-related regulations and the code of conduct adopted by IAEA. It will also be necessary to have available enough trained personnel to take care of areas like the calibration and operation of radiation detectors, data analysis, and decision-making concerning NORM.

In the United States, the NCRP recommended in 2002 that the EPA, the Nuclear Regulatory Commission (NRC), and the Department of Energy (DOE), in agreement with state regulators, develop a system to control the recycling of TENORM-contaminated scrap metal, produced either domestically or abroad (NCRP, 2002).

Spain has achieved an exemplary success by adopting radiological control of metal scrap on a voluntary basis (Rodríguez Martínez, 2008). The Spanish experience involves collaboration between the government and industry, as well as sharing costs among all collaborating entities (Rodríguez Martínez, 2008). The Spanish approach deserves to be followed closely to check the feasibility of adopting a similar model at the international level.

CHAPTER 4

Cosmic Radiation, Including its Effects on Airline Crew, Frequent Flyers, and Space Travel

4.1. The Issue

The growing mobility of our society is associated with a continuing increase of air travel. This leads to an increase in the number of persons exposed to ionizing radiation called *cosmic radiation*, as pilots and cabin crews, but also of frequent travelers and, to a lesser extent, as passengers. Airline promotion programs encourage this trend, for example, offering special bonuses for passengers traveling more than 100,000 miles per year. Furthermore, astronauts are exposed to cosmic radiation during their missions. In addition to occupational and non-occupational radiation exposure, cosmic radiation also impacts microelectronic components onboard of aircraft, spacecraft, and space stations. Since 1990, this type of radiation exposure has become a subject of concern to the International Commission on Radiological Protection (ICRP).

As a matter of fact, the possibility of developing in the mid-1970s a large-scale supersonic air transport (SST) for passengers and cargo, to be used in the subsequent decade, raised the interest of some scientists to a number of questions related to the index of absorption dose rates for passengers and crew exposed to cosmic radiation at high altitudes. Measurements carried out at several altitudes and geomagnetic latitudes with a dose-equivalent device developed at Brookhaven National Laboratory (BNL), in the United States, indicated a constant increase in the radiation dose between geomagnetic latitudes 36°N and 58°N at a rate of about $0.37\,\text{nGy}\,\text{h}^{-1}$ from 3 to 7 km altitude. From 7 to 10 km altitude, the radiation dose increase at latitude 36°N was 0.78 nGy, but with a positive gradient of $1.7 \times 10^{-2}\,\text{nGy}\,\text{h}^{-1}\,(°\text{Lat})^{-1}$ for the North Hemisphere latitudes mentioned above (ACRBA-SST, 1975).

4.2. Source Term: Cosmic Radiation

It is interesting to remember here that in 1901, Charles Thomson Rees Wilson had stated that as a consequence of radiation coming from

Radioactivity in the Environment, Volume 17
ISSN 1569-4860, DOI 10.1016/S1569-4860(09)01704-5

sources beyond the Earth's atmosphere, radiation like the Röntgen or cathode rays might have penetration power much higher than anticipated thus far (Wilson, 1901). Not long after, investigators from the University of Toronto and from McGill University, both in Canada, demonstrated that 5 cm of lead would decrease by 30% the intensity of the mysterious radiation coming from the skies (McLennan and Burton, 1902; Rutherford and Cooke, 1903; see also Parizot, 2007). However, five additional tons of lead would not reduce the intensity of the radiation just mentioned. In the next three years nothing significant was added to knowledge of the highly energetic radiation, until 1907, when the priest Theodore Wulf, from the Saint Ignatius College, in Valkenburg, the Netherlands, used an electroscope and an ionization chamber to measure radiation that he believed was coming from the surrounding soil. The priest started not to trust his own instruments when measurements made at the top of the Eiffel Tower indicated that the ionization rate was much higher than he had expected (Wulf, 1910; Ziegler, 1989).

Figure 4.1 illustrates in a graph the values expected and measured by Wulf. He distrusted his instruments because he was unable to understand, to say nothing about explaining, what was happening. A factor of almost 9 in the ratio measured ion density to expected ion density at about 300 m from the base of the Eiffel Tower asked for a convincing scientific explanation.

After radon gas was discovered in 1900 by Friedrich Ernst Dorn (Dorn, 1900), a number of experiments were performed with the newly discovered gas. In Wulf's experiment, the exponential curve shown in Figure 4.1 was not only expected but also compatible with knowledge about gas behavior at that time. There is speculation, however, that Wulf at the time was suspicious that there was radiation coming from extraterrestrial space. But it was Albert Gockel who, in 1910, decided to put Wulf's instruments in a balloon, which was raised up to 4.5 km, after having made radiation measurements in the following mountain peaks (Gockel, 1909, 1911, 1912): Rothorn (2,300 m), Zermatt (2,600 m), and Matterhorn (3,000–3,300 m). The measurements made at the mountain peaks convinced Gockel that there should be radiation coming from space. It was then that Millikan stated that the important fact is that the instrument readings do not go to zero above 1,000 m height, as one should expect if the radiation sources were in the Earth (Millikan, 1911, 1913, 1993).

In 1912, a new environmental radiation component was discovered during a series of balloon flights at heights up to 5,000 m above ground: contrary to expectations, its intensity *increased* with increasing distance from the sources of natural background radiation known hitherto, that is, natural radionuclides contained in soil and rocks on the ground (Hess, 1912a, 1912b). The Austrian physicist, Dr. Victor Hess (1883–1965), observed that the radiation measured in a balloon decreased gradually up to 700 m, but when the balloon reached 1.5 km the radiation started to increase again,

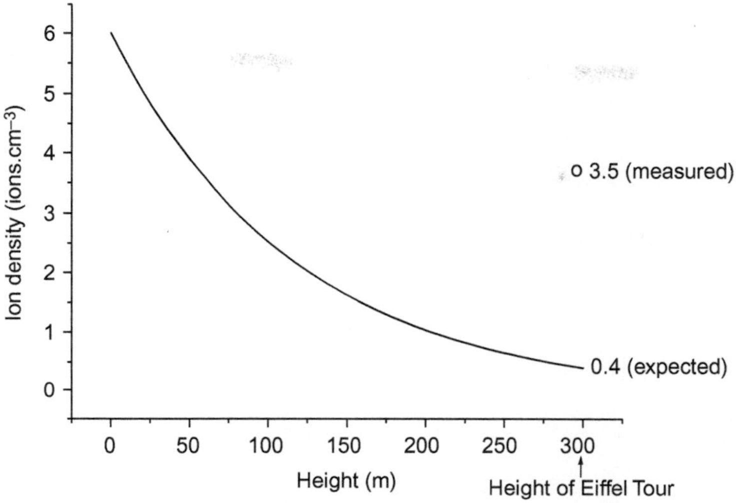

Figure 4.1 Graph of the values expected by Wulf plus the measured value at 300 m height at the Eiffel Tower.

and at about 5 km the radiation had an intensity which was twice that measured while the balloon was still on the soil (Hess, 1912a, 1912b; O'Brien and Skalski, 1996). Victor Hess received the Nobel Prize in 1936 for his discovery. This penetrating radiation originates outside the earth and is called *cosmic radiation*. The sources of cosmic radiation are galaxies in outer space and the Sun in our solar system. Cosmic radiation impact on Earth has two main components: primary galactic cosmic radiation (GCR), which normally is dominant, and a component from the Sun from solar particle events (SPEs). SPEs follow a cycle of about 80–100 years (*Gleissberg cycle*). During solar maximum conditions sunspots are more frequent, inducing the probability of coronal mass ejections. However, only a small fraction of SPEs, on average one per year, produces large numbers of high-energy protons, which cause an observable increased intensity in cosmic radiation fields at aviation altitudes, and can be observed with neutron monitors at ground level (ground level events, GLEs).

GCR is considered to be the result of the acceleration of ambient interstellar matter to high energies by the shock waves of exploding supernovae. During its passage through space, the composition of GCR is modified by interaction with interstellar matter, as illustrated in Figure 4.2.

GCR, originating from sources outside our solar system, is made up of about 98% baryons and 2% electrons (Reitz, 1993). It consists mainly of protons (87% of the baryons) and to a lesser extent of helium ions (11%) and heavier ions (ranging from carbon to iron; 1%), with energies from 10^8 eV to more than 10^{20} eV. It is assumed that they originate from large stellar explosions or the fast rotating neutron stars resulting from them.

Figure 4.2 After, and before, Supernova 1987 (reproduced with permission from Anglo-Australian observatory/David Malin Images).

The other component of the GCR is due to *solar flares*, sudden eruptions of the chromosphere of the Sun, ejecting charged particles with energies mostly below 10^8 eV up to several 10^{10} eV. Such events occur most frequently at the end of maximum solar activity (Foelsche et al., 1974).

Solar flares frequently cause a magnetic storm on Earth, depressing its magnetic field and allowing particles of lower energy to penetrate into the magnetosphere. There are also short-term perturbations in the Sun's magnetic field which can result in increased solar activity, large solar flares, and coronal mass ejections. These modulations follow the 22-year solar activity cycle, which consists of two 11-year cycles, separated by the reoccurring reversal of the Sun's magnetic field polarity. The changes of the Sun's magnetic field interact with the magnetic field of the Earth, leading to decreased intensities of GCR (*Forbush decreases*) (Forbush, 1937). The magnitude of the Forbush decreases is highly variable, depending on the stage of the solar cycle. When solar sunspot activity is at a maximum during this short cycle (i.e., about every 11 years), the increased solar field acts to screen out low-energy GCRs. Therefore cosmic radiation intensities vary in a cyclical pattern, anti-coincident with the solar activity. Using Deep

River cosmic ray neutron monitor data (O'Brien, 1973), the relationship between the Aa index from 1868 to 1967 – and derived from readings made at antipodal stations in Greenwich and Melbourne, or their substitutes – and the modulating potential radiocarbon records from 1005 to 1860 (Stuiver and Quay, 1980), O'Brien and colleagues were able to reconstruct the variations of the heliocentric potential over the last millennium showing conspicuously the Wolf, Sporer and Maunder minima (O'Brien et al., 1996).

The development of new computational models of cosmogenic isotope production in the atmosphere allowed the reconstruction of long-term solar activity (Masarik and Beer, 1999; Beer, 2000). Multimillenia reconstructions of solar activity throughout the Holocene have been made using ^{14}C and geomagnetic data (Yang et al., 2000; Usoskin et al., 2007; Usoskin, 2008) (Figure 4.3).

Figure 4.4 shows the registration of the classical Forbush decrease of October 28–30, 2003 by the *Milagro* TeV gamma-ray ground level telescope, which is sensitive to solar energetic particles (Ryan and the Milagro Collaboration, 2005). The *Milagro* instrument is located in the Jemez mountains at an altitude of 2,630 m in northern New Mexico. It detects solar and galactic protons by means of electromagnetic (hadronic) showers entering a one-acre pond of water. The photomultipliers (PHTs) are submerged in water, so the Cerenkov light from relativistic muons is detected by one or several PHTs. The upper and lower graphs show the responses of the Cerenkov and high threshold scaler for galactic cosmic ray

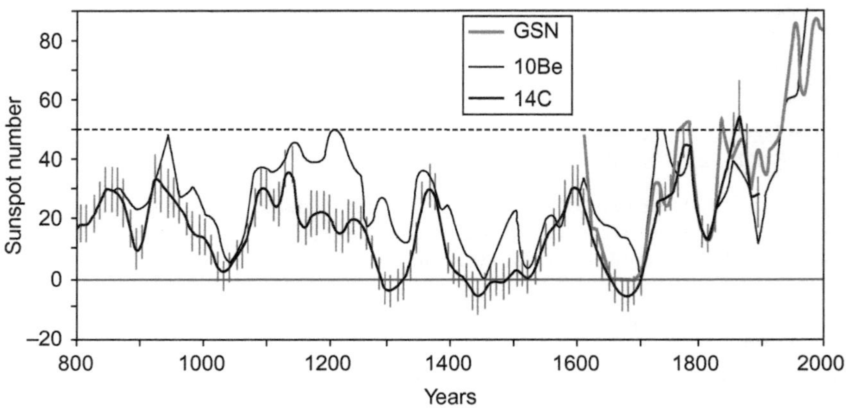

Figure 4.3 Sunspot numbers averaged at 10-year intervals: in the upper box GSN means actual group sunspot number; 10Be represents the reconstruction based on ^{10}Be (Usoskin et al., 2007; Usoskin, 2008); and 14C line with error bars represents the reconstruction based on ^{14}C. The horizontal dotted line at sunspot number 50 depicts the high activity threshold. This picture is reproduced with permission of Dr. Ilya Usoskin.

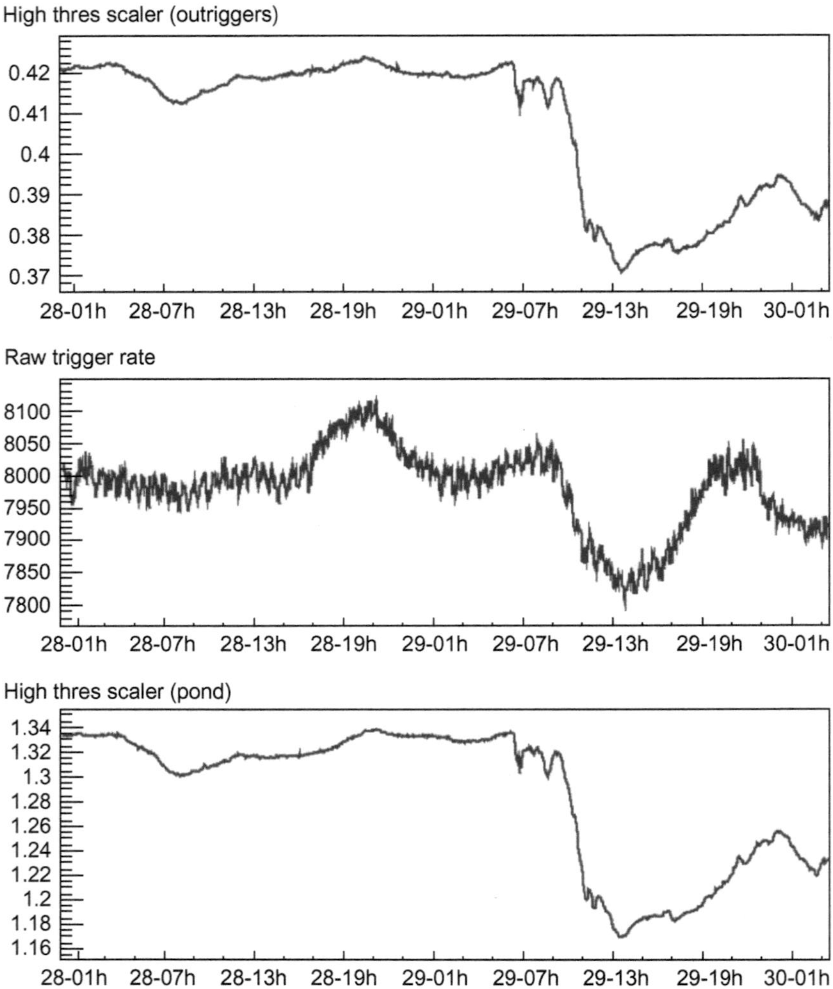

Figure 4.4 Measurement of the classical Forbush decrease (Ryan and the Milagro Collaboration, 2005). Reproduced with special permission from Dr. James Ryan. The authors are grateful to Dr. James Ryan for allowing these graphs to be reproduced.

proton detectors, respectively. The threshold for an event trigger is of the order of 50 GeV. The graph in the middle shows the raw trigger rate in the pond (Ryan and the Milagro Collaboration, 2005). The Forbush decrease is clearly shown in all the three graphs.

The GCRs that are not deflected by the solar magnetic field now encounter the magnetic field of the Earth. The result of the cosmic radiation hitting the Earth is a bombardment of its magnetosphere by a nearly isotropic flux of charged particles with different energies. However, only part of the cosmic radiation actually reaches the surface of the Earth. The charged particles cause pressure on the Earth's magnetic field on the side exposed to the Sun and produce a magnetospheric tail at the rear. While this perturbation is considerable at high altitudes, the dipolar structure of the magnetic field predominates at altitudes below a few Earth radii (a radius is equal to 6,370 km) (Bottolier-Depois et al., 2000). Therefore, the magnetic field offers less resistance to incoming particles at the polar cones, centered at (or near) the magnetic poles. Charged particles are deflected by the component of the magnetic field that is perpendicular to the direction of particle motion. This means that the GCR is deflected more at the equator than near the poles. Since GCR and solar particles with lower energy are deflected by the magnetic field of the Earth, only charged particles with high energies are able to penetrate this magnetic field and enter the Earth's atmosphere. Figure 4.5 shows the cosmic ray flux as a function of particle energy.

Since access is easier at the poles than at the equator, this produces a geomagnetic latitude effect for cosmic radiation. It is mostly the high-energy protons which interact with the atoms and molecules in the atmosphere (mainly nitrogen and oxygen), resulting in a cascade of interactions and reaction products (*secondary particles*; Figure 4.6 – see Zombeck, 2006).

Upon entry into the Earth's magnetic field the charged particles are deflected by the Lorenz force in which their trajectories have a curvature of radius *r*:

$$r = \frac{R_\mathrm{p}}{B_\mathrm{p}} \tag{4.1}$$

where B_p is the component of the Earth's magnetic field perpendicular to the direction of the motion, R_p the *magnetic rigidity* (*MR*), which is related to the particle's momentum p and charge q.

The R_p of a particle is the quantity which describes its penetrating power, and is defined as the product of the magnetic field strength component perpendicular to the particle momentum and the radius of curvature of the particle trajectory, equal to the particle momentum divided by particle charge (q).

$$R_p = \frac{pc}{q} \tag{4.2}$$

where c is the speed of light (i.e., 300,000 km s^{-1})

The quantity R_p is usually expressed in units of GV. The *cutoff rigidity* is the minimum rigidity for a particle to penetrate to a location on Earth. At the pole the magnetic field lines are nearly vertical, R_p equaling 0 GV; at the equator magnetic field lines are parallel to the surface of the Earth

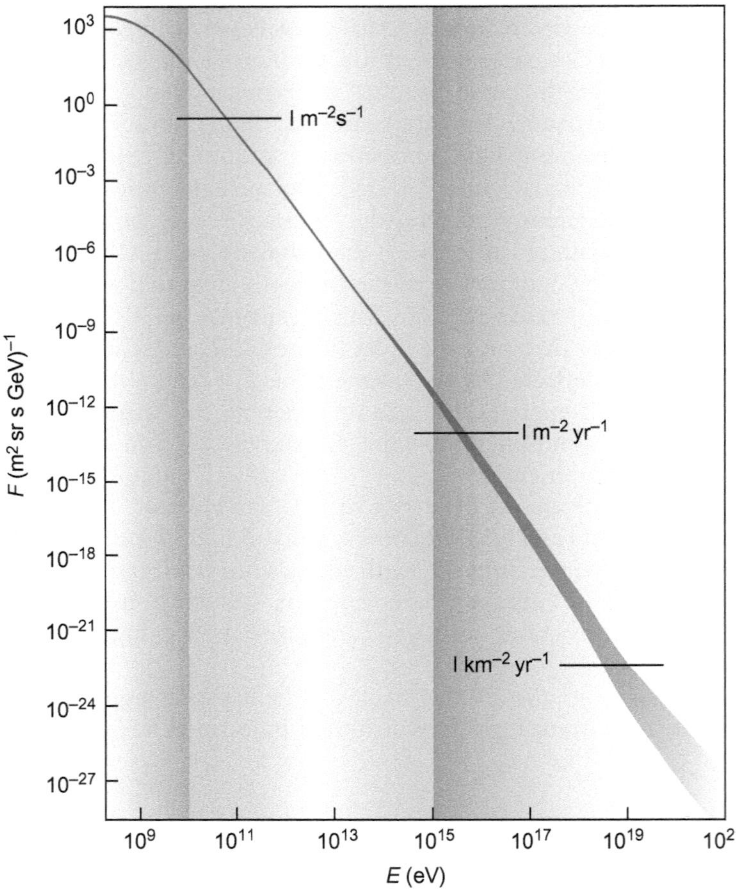

Figure 4.5 Graph of the cosmic ray flux as a function of the particle energy (solar, intermediate, and galactic plus extragalactic). Reproduced from Wikipedia Free Encyclopedia under the *GNU General Public License – Free Documentation License.* http://en.wikipedia.org/wiki/File:Cosmic_ray_flux_versus_particle_energy.svg. Last visited on July 15, 2009. See also Swordy (2001).

with R_p about 15 GV. There is no maximum for the primary protons at high latitudes, but at low latitudes the high cutoff rigidity of the Earth's magnetic field removes a significant fraction of low-energy primaries (Lewis et al., 2001)

The high-energy particles incident on the atmosphere interact with nitrogen and oxygen nuclei producing lower Z nuclei, secondary protons, neutrons, and charged and uncharged pions (Bartlett, 2004). The secondary protons and neutrons generate more nucleons – the *hadron cascade*. The neutron component dominates the hadronic cascade at lower altitudes as a

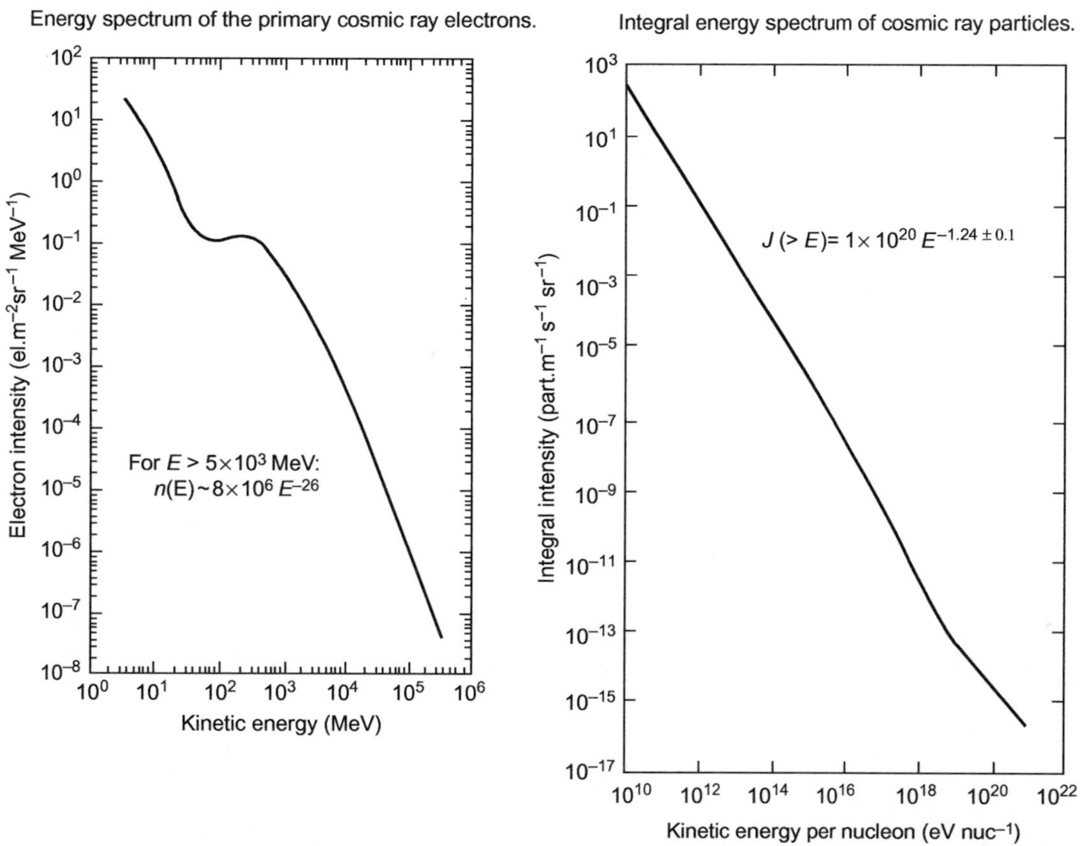

Figure 4.6 Different components of the secondary particles produced by an incident primary particle (Zombeck, 2006). See also NASA

result of their longer mean free path. The neutral pions decay into high-energy photons, which produce electron–positron pairs leading to the production of annihilation photons thence pair production and so on – the *electron–photon cascade*. The decay of the charged pions generates muons – the *muon cascade*. Electrons and positrons from the decay of muons contribute to the electron–photon cascade.

The creation of these secondary particles is in competition with attenuation by the atmosphere (i.e., cosmic radiation varies with altitude). It increases with increasing altitude as the shielding of the atmosphere decreases and reaches a maximum at about 20 km (*Pfotzer maximum*). Due to the cutoff rigidity curve displaying the so-called *geomagnetic knee*, which is a fairly large region above approximately 50°N in Canada or 70°N in Siberia, cosmic radiation levels are constant with increasing latitude. Above the Pfotzer maximum the ionization decreases and reaches a constant value at and above 50 km.

Muons are produced in the high atmosphere by means of nuclear reactions between PCRs and nuclei of gaseous molecules. When high-energy protons, which are the majority of RCRs, reach the upper layers of the atmosphere, a large number of nuclear interactions occur. As a result of these interactions the SCRs are created. The majority of the SCRs are pions with very short half-life. The pions occur in three types: π^0 (neutral), π^+, and *e* π^-. The latter two are the particle and antiparticle of each other. While the π^0 produces two protons in its decay process, which in turn produce other particles, the two charged pions decay as follows:

$$\pi^+ \rightarrow \mu^+ + \nu_\mu \quad \text{producing an anti-moun and a muon-neutrino} \tag{4.3}$$

$$\pi^- \rightarrow \mu^+ + \bar{\nu}_\mu \quad \text{producing a muon and a muon-anti-neutrino} \tag{4.4}$$

Relativistic muons have two important characteristics. First, charged muons have half-lives of the order of 2×10^{-6} s, thus much longer than those of pions (π^0, e.g., has a half-life of only 8×10^{-15} s). Second, charged muons easily reach the Earth's surface, because they do not undergo many interactions. As a consequence, charged muons constitute almost the totality of the hadronic shower that reaches the soil, contributing significantly to the natural external background radiation.

The determination of the rate of exposure to external radiation depends on a measurement of ionizing radiation by some kind of instrument. Proportional counters, for example, are often used to determine the rate of exposure to external radiation. The measurement of the rate of exposure to external radiation depends on the following factors:

1. latitude;
2. altitude;

3. emission of radiation by natural radionuclides of the radioactive series plus ^{40}K; and
4. cloud coverage of the skies.

The first three factors have been studied for many years now. Back in 1949, Victor Hess was already worried about the biological effects of cosmic rays (Hess and Eugster, 1949). In the mid-1960s, scientists of the Health and Safety Laboratory (HASL) in New York maintained a monitoring program to measure continuously the ionizations caused by cosmic rays in the lower atmosphere (Beck and Lowder, 1966). At the end of the 1980s, André Bouville and Wayne Lowder called the attention of the radioprotection international community to the fact that it would be difficult to estimate the long-term doses to human populations due to cosmic radiation with a high degree of certainty, because there were significant spatial variations as a function of altitude and latitude (Bouville and Lowder, 1988). At the beginning of this century, John C. Gosse and Fred M. Phillips emphasized that the *in situ* rate of production of radiogenic radionuclides increased with increases in altitude and latitude (Gosse and Phillips, 2001). More recently it was verified that when monitoring the SCRs there might be variations of up to 15% in measurements made at sea level due to the influence of the barometric pressure, solar activity, and temperature. In 1935, the variation of the detected SCR at sea level with the barometric pressure was known as the *barometric paradox* (Clay, 1935). Only recently an explanation was proposed for the *barometric paradox* based on the increasing number of interactions among the particles produced in the hadronic shower (de Freitas et al., 2007). An increase in the number of interactions occurs in the hadronic shower with the increase of atmospheric pressure, because it decreases the mean free path of the particles. As a consequence, when the hadronic shower reaches sea level it is enlarged, a smaller number of particles reaching the detector which maintains the same geometric configuration (de Freitas et al., 2007). Figure 4.7 illustrates a hadronic shower originated by a high-energy proton. If a detector is represented by the black bar at the bottom of Figure 4.7, the enlargement of the hadronic shower would result in a smaller number of particles reaching that detector. A correction factor for this effect could be calculated as a function of the atmospheric pressure.

The muons at the base of a hadronic shower have a higher percentage composition in the external radiation measured offshore, far from the influence of naturally occurring radionuclides extant in soil and rocks. Inland the geological configuration underneath the sites and the radon gas concentration near the detector contribute significantly to the measured external radiation, because of the gamma emissions of the ^{238}U and ^{232}Th natural series, and the 1.46 MeV gamma-ray line from ^{40}K. In particular, the gamma-ray lines of ^{214}Bi, which is a member of the progeny of ^{222}Rn,

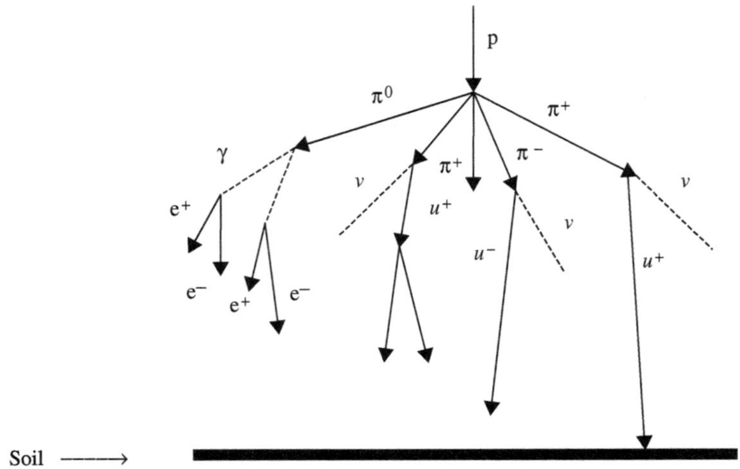

Figure 4.7 Schematic representation of a hadronic shower originated by a high-energy proton.

increase the number of events registered by detectors sensitive to gamma radiation at relevant energies for those emissions.

As was shown in Table 3.5, dose coefficients to convert soil activity concentrations (Bq kg^{-1}) to absorbed dose rates (nGy h^{-1}) from three distinct sources have slightly different values. These dose coefficients are widely used throughout the world.

On the other hand, the dose coefficients used for cosmic radiation are obtained through sophisticated computational codes, which have evolved significantly during the last three decades. The fundamental mathematics and physics that allowed the development of techniques used for cosmic radiation were published in a single volume in 1980 (Nelson and Jenkinns, 1980). This book describes in a didactic way the fundamentals of the initial codes used to describe the hadronic showers to be used in the following: (i) shielding particle accelerators; (ii) to evaluate the effects of cosmic radiation; (iii) to estimate the radiation doses received by crew of commercial flights, by military pilots, and with some modifications by astronauts in several missions; and (iv) in other applications in medicine and dosimetry. Table 4.1 lists the main codes used before 1980 to describe hadronic showers.

Validation of the codes listed in Table 4.1 was not easy, it implies either measurements, or even more sophisticated codes. By and large, the instruments used on board airplanes and space ships are not able to detect all particles used in the codes. In addition, the dose dependence on particle energy makes the problem even more complex. All codes listed in Table 4.1 used Monte Carlo techniques, with the sole exception of LUIN. All of them have been improved with time. Other codes have been created to describe hadronic showers with higher or lesser degree of sophistication.

Table 4.1. Computational codes used before 1980 to describe hadronic showers.

Code name	Author	Method	Model	Reference
LUIN	O'Brien	Analytical	Empirical	O'Brien (1973)
CASIN	Van Ginneken	Monte Carlo	Empirical	Van Ginneken (1975)
FLUKA, KASPRO	Ranft	Monte Carlo	Empirical	Ranft (1972)
Dubna Code	Barashenkov et al.	Monte Carlo	Intranuclear cascade	Barashenkov et al. (1972)
HETC	Armstrong et al.	Monte Carlo	Intranuclear cascade	Armstrong and Chandler (1972)

However, when one wants to evaluate radiation in space during a solar event, it is easier to validate a model by using a series of instruments which are adequate to measure high fluxes of particles during the event (Clucas et al., 2005).

It is worth mentioning here that the basic physics and mathematics to develop computational codes to describe hadronic showers did not change significantly during the last decades. The more important atomic interactions are ionizing and excitation effects, and in some cases coulomb scattering. The elastic interactions of hadron–nuclei have a tendency to increase the base of the hadronic cone as the altitude diminishes, but the associated uncertainties are large.

It is beyond the scope of this book to discuss further on the interaction between hadrons and matter, or coupling of the hadronic showers. In any case, it is worth mentioning here that knowledge of the basic phenomenology associated with the interaction of hadrons (i.e., strongly interacting particles) with matter is important to understanding the formation of a hadronic shower. From the radioprotection point of view, however, the components of the cosmic radiation spectra which contribute to the measurement of external radiation rate are those ionizing directly (muons and photons) and indirectly (neutrons). In order to establish what percentage contribution cosmic radiation makes to an external radiation rate measurement, it is necessary to measure those components at sea level, and at relevant altitudes. Recently Keran O'Brien made a series of comparisons between calculated and measured values of ionization and neutron profiles due to cosmic radiation, and also for muons detected in underground laboratories (O'Brien, 2005, 2008).

The cosmic component measured in the external radiation is not included, in general, in the conversion to exposure rate at sea level. Dose rate estimates due to cosmic radiation to persons flying in commercial airplanes at altitudes between 9,000 and 12,000 m vary with latitude and altitude. Lower doses are received along the equator, while larger doses are received at latitudes above 50°N. Results from measurements made by the European Community indicate that crews of commercial airplanes received dose rates in the range between 1.55 μSv h^{-1} – near the equator at 9,500 m during the maximum of solar activity (solar max) of 1991 – and 9.5 μSv h^{-1} – at latitudes above 50°N, at 12,000 m, during the minimum solar activity between 1974 and 1976 (EC, 1996a, 1996b).

Some years ago there was an effort to improve the conversion coefficients for use in cosmic ray dosimetry (ICRP 74, 1997a, 1997b; Ferrari and Pelliccioni, 2003; ICRP 92, 2004). The main point in establishing dose coefficients for cosmic radiation is, however, the experimental determination of the muon component of the external radiation.

The percent composition of each component of external radiation due to cosmic rays constitutes important information, at least as far as radioprotection is concerned. Table 4.2 summarizes preliminary estimates made in Japan on the percent composition of the components of external radiation due to cosmic rays.

The need to determine separately the dose conversion factor for each component of external radiation becomes evident from the data presented in Table 4.2. The muon component of the hadronic shower has a significant percentage in the composition of external radiation when measurements are made offshore, far from the influence of natural radionuclides. When a comparison of inland and offshore external radiation measurements was made with the same instrument, a reduction of only 20% was observed offshore, which could be interpreted as a major muon component of external radiation (de Freitas et al., 2007). In any case, the actual contribution of muons to the external radiation is still to be determined in a manner acceptable by the radioprotection community.

The cosmic ray interactions result also in *cosmogenic radionuclides* (e.g., ^{3}H and ^{14}C), which are naturally radioactive. As a result of these interactions, alpha particles, electrons, muons, neutrons, pions (charged and neutral), positrons, protons, and gamma radiation are formed in the air.

The interaction of primary particles with the elements in the atmosphere and the magnetosphere of the Earth is the reason for the *Van Allen radiation belts* (James Alfred Van Allen, 1914–2006) encircling the

Table 4.2. Summary of the percentage of each component of the annual exposure to external radiation (μSv y^{-1}) due to cosmic rays (Rasonlonjatovo et al., 2002).

Type of radiation (target)	%
Gamma	8
Muon (tissue)	47
Muon (air)	40
Neutron	5

Earth. The Van Allen radiation belt is named after its discoverer James Van Allen from the University of Iowa. The discovery was a consequence of Van Allen's analysis of the data obtained with Geiger counters on board the satellites Explorer 1, which had been launched on January 31, 1958, and Explorer 3, launched on March 26, 1958 (Van Allen et al., 1958; Van Allen and Frank, 1959; Kip, 1962; see also Ludwig, 2004). It is fairly well understood that the inner and outer Van Allen belts result from different processes. The inner belt, which is the one actually discovered by James A. Van Allen, consists mostly of energetic protons, and is the product of the decay of albedo neutrons which are themselves the result of cosmic ray collisions in the upper atmosphere. The outer belt consists mainly of electrons.

The Van Allen belts of charged particles correspond to a drop in pressure of the magnetic field in which the charged particles are trapped. In general, only two belts are consider at about 700–60,000 km above the Equator.

In 1990, the Combined Release and Radiation Effects Satellite (CRRES) discovered a third radiation belt located between the inner and outer Van Allen belts (see, e.g., Baker, 1994). Also, around May 8, 1998, there was a series of large, solar disturbances that caused a new radiation belt to form in the so-called "slot region" between the inner and outer Van Allen belts. The new belt eventually disappeared once the solar activity subsided.

The Earth has actually two permanent radiation belts. Each one has a different origin. The inner belt, the one discovered by Van Allen's Geiger counters, occupies a compact region above the equator, and is a by-product of cosmic radiation. This inner belt is populated by protons of energies in the 10–100 MeV range, which readily penetrate spacecraft and which can, on prolonged exposure, damage instruments and be a hazard to astronauts. The outer belt is thought today to consist of plasma trapped by the Earth's magnetosphere. The outer belt is larger and more diffuse than the inner one. The outer belt is populated mostly by electrons, but there are also various ions like energetic protons, and a smaller percentage of alpha particles plus oxygen ions.

Particles present in the radiation belts can penetrate space ships and satellites damaging equipment and astronauts. Radiation belts can be amplified by high-altitude nuclear explosions, like the Starfish Prime test on July 9, 1962 (Johnston, 2006), or can even be created temporarily just after a solar magnetic storm (Baker et al., 2004; Horne et al., 2005). Radiation doses up to 10 Gy could be received in a worst case scenario, such as traveling in space during one of the four major solar storms in the last three 11-year solar cycles in the low-Earth satellite orbits used to cross the South Atlantic Anomaly (SAA), a region over the coast of Brazil, where the inner radiation belt reaches down to altitudes of the order of 200 km.

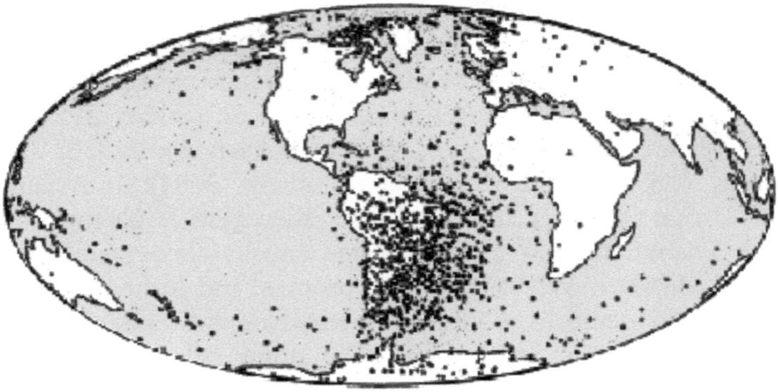

Figure 4.8 Locations of upset events suffered by LEO satellite (UOSAT-2). *Credits*: Margaret Ann Shea, Geophysics Directorate, Philips Laboratory.

The low-latitude satellites in low-altitude earth orbit (LEO satellites) suffered a large number of upset events, as one can see in Figure 4.8.

Cornelius A. Tobias (1918–2000) had predicted, back in 1952, before the discovery of the Van Allen belts that there would be a radiation danger to astronauts trying to explore space (Tobias and Todd, 1974). As a matter of fact, it has been reported that light flashes and streaks were observed randomly by astronauts of Apollo 11–15, due to interaction of energetic heavy PCRs with the retina (Budinger et al., 1972).

Astronauts embarking on or returning from journeys to the Moon or Mars will need to cross the Van Allen belts, and thus will have to be exposed to high levels of radiation, even if only for short periods (see, e.g., NRC, 2006). Both manned and unmanned spaceflights tend to stay out of this region to avoid very high radiation doses (Figure 4.9).

4.3. Doses Due to Cosmic Radiation

Man is exposed to cosmic radiation in several ways:

On the *surface* of the Earth people are largely shielded by the atmosphere. The effect of this shielding is equivalent to about 4 m of concrete. The main source of ground exposure is due to muons, formed from charged mesons because of their long mean free path in the atmosphere (energies typically range from 1 to 20 GeV) contributing about 80% of the absorbed dose rate in free air from the directly ionizing radiation. The remainder results from electrons (UNSCEAR, 2000). At lower levels of the atmosphere and at sea level, the dependence on the 11-year solar cycle is small compared to the uncertainty in these estimates. Due to the

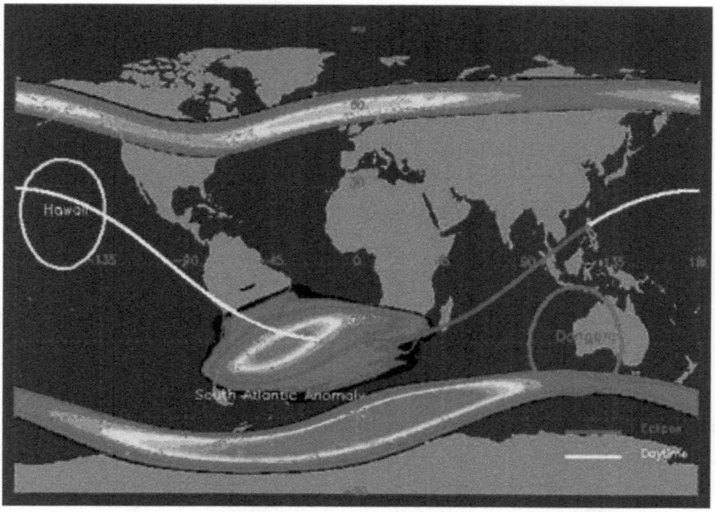

Figure 4.9 GALEX orbit image of the South Atlantic Anomaly (SAA). The particle flux associated with the SAA is modeled (dark gray = low, light gray = high). This picture was reproduced from the site www.galex.caltech.edu/researcher/techdoc-ch2.html.

Earth's magnetosphere and the atmosphere, a powerful shield protects people living on the ground from cosmic rays which would otherwise cause an annual dose in excess of 1 Sv (Bottolier-Depois et al., 2000)

The radiation field found at *jet aircraft altitudes* (typically selected between 6 and 12 km in order to save fuel; in the case of supersonic flights up to 18 km) results mainly from the interaction of GCR particles and the atmosphere. The radiation field inside an aircraft flight is further modified by the structure and materials of the aircraft and consists of the following main components: neutrons, electrons, positrons, protons, photons, and muons.

Space travelers are exposed to an even more complex radiation environment, such as in the Van Allen belts, galactic cosmic rays, SPEs, and neutrons produced by the interaction of the high energy protons with the shielding of the space vehicles and the tissue of the crew members.

Exposure to cosmic rays on the ground is basically a personal decision on where to take up residence. Since it cannot be influenced by technology in any reasonable manner, it therefore is not subject to any control. Based on an ion-pair production rate of $2.1\,cm^{-3}\,s^{-1}$, this results in a global average effective dose rate outdoors of $32\,nGy\,h^{-1}$ at sea level; indoors, with a mean shielding factor of 0.8, this equals an effective dose rate of $26\,nSv\,h^{-1}$. Assuming that on average about 80% of the time is spent indoors, the global average effective dose from the ionizing component of cosmic radiation at sea level is estimated to be 240 μSv (UNSCEAR, 2000). Depending on the place of residence, the total annual effective dose from cosmic and cosmogenic radiation ranges from 0.3 to 1.0 mSv with an average of 0.39 mSv (UNSCEAR, 2000).

However, the cosmic ray exposure of crew and passengers in civil aviation, as well as of space travelers, represents a case of technologically enhanced natural radiation and will be discussed in more detail below.

4.3.1. Radiation exposure of crew and passengers in civil aviation

Crew and passengers receive an elevated exposure to cosmic radiation as compared to residents on the surface of the Earth. The actual level of exposure depends on: (a) the GCR component, (b) the SPE component, (c) the secondary particles, (d) the structure of the aircraft, and (e) content of the aircraft. The actual value depends on the path taken. Since galactic radiation entering the Earth's atmosphere varies inversely with the cycles of solar activity, the solar wind intensity (*heliocentric potential*) is monitored regularly.[1]

The shielding provided by the atmosphere at a given altitude is determined by the *atmospheric depth*, that is, the mass thickness of the air above. At civil subsonic flight levels, the residual atmosphere measures about 190–270 g cm^{-2}; at this altitude radiation exposure is caused mainly by secondary particles. Representative values of the main components of ambient dose equivalent at aircraft altitudes and temperate latitudes are as follows (Bartlett, 2004):

Neutrons	55%
Electrons and positrons	20%
Protons	15%
Photons	**5%**
Muons	5%

There is not a significant contribution to dose equivalent from energetic primary heavy charged particles (HZE) or fragments (O'Sullivan et al., 1999).

For the determination of the occupational exposure of an aircraft crew to cosmic radiation, it is necessary to determine which dose quantity is relevant and how to define the exposure time to cosmic radiation (flight hours) (Bartlett, 1999a, 1999b). Survey meters permit the determination of the *ambient dose equivalent*, which is used together with *radiation weighting*

[1] Monitoring is carried out by the Solar and Heliospheric Observatory (SOHO spacecraft; details at www.soho.nascom.nasa.gov).

factors (w_R). Radiation protection foresees the assessment of the *effective dose* and associated *quality factors* (Q_F) in order to limit the potential detriment resulting from the exposure to ionizing radiation; w_R and Q_F are defined differently. Since effective dose cannot be measured directly, the operational quantity of interest is the ambient dose equivalent $H^*(10)$ – defined as the dose equivalent in a 10 mm depth of the ICRU sphere in an aligned and expanded radiation field.

In Figure 4.10, the values for Q_F are compared for different occupationally exposed population groups.

A

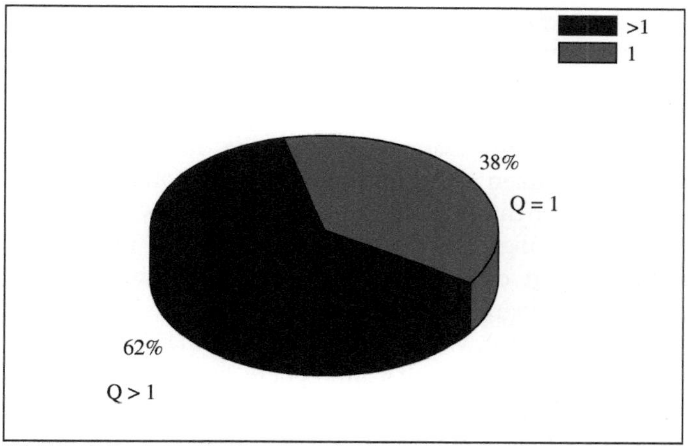

Aircrew on board a trans-atlantic flight (~11 km altitude).

B

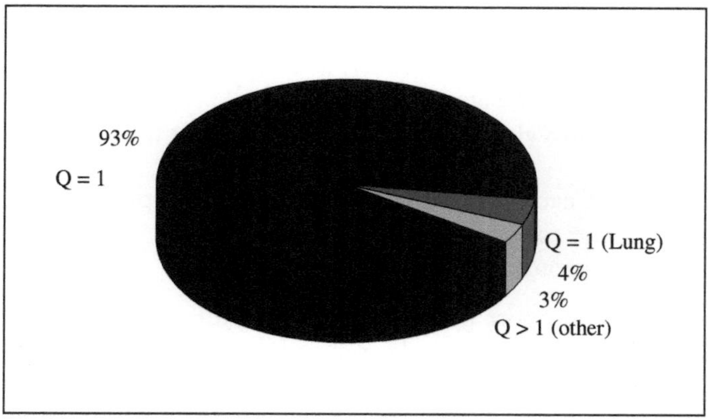

US atomic radiation workers.

Figure 4.10 Comparison of quality factor Q_F for (a) aircrew on board a trans-Atlantic flight (~ 11 km altitude) and (b) ground-based US radiation workers. Graphs adapted from Wilson (1998), and Lewis et al. (2001).

The magnitudes of ambient dose equivalent and effective dose are comparable, except for the proton component of the radiation field inside the aircraft, for which the magnitude of the effective dose is approximately 500% higher than that of the ambient dose equivalent (Pellicioni, 1998). Therefore, the values of the effective dose are numerically greater than the ambient dose equivalent: up to 30% for cruising altitudes of 9–12 km and about 50% at altitudes of about 18 km.

The exposure level on board an aircraft can either be measured or calculated:

a) In-flight measurements have been made with a wide range of different devices, with good agreement of measurement, such as TLD, bubble detector, etched track detector, fission foil with damage track detector, electronic dosimeter, recombination ionization chamber, multisphere spectrometer, silicon-based detector, scintillation detector, semiconductor spectrometer, and tissue equivalent proportional counter (TEPC). TEPC approximates reasonably well the relevant operational dose quantity, the ambient dose equivalent. It is important that these devices are calibrated for the response to neutrons (energy, angle dependence).
b) Alternatively, in-flight dose rates and route doses can be calculated by radiation transport codes for a given altitude (pressure rather than radar altitude) and geographic location (geomagnetic latitude rather than geographical location), which use as input parameters the cosmic radiation field at the top of the Earth's atmosphere, speed, and flight path of the aircraft. Flight duration for pilots and cabin crew are calculated, accounting for the whole period starting with the departure from the terminal, taxiing on the runway, ascending during take-off and approaching operational altitude, time spent at the cruising altitude, descending during the landing phase, and taxiing to the terminal at the destination (*block hours*). The actual annual flying time of staff members at different airlines ranges from about 300 to 900 h y^{-1} (average: about 500 h).

   These programs also account for the following modifying parameters:

1. interactions of each particle with the various components of the air;
2. variation in the magnetic field of the Sun; and
3. count rate of ground level monitors.

The various transport codes are based on different modeling assumptions. The LUIN transport code is based on an analytic solution of the Boltzman transport equation (O'Brien, 1978). Using the computer code LUIN, the US-software CARI was developed for the calculation of route doses, incorporating an advanced geomagnetic cutoff rigidity model and PCR spectrum (O'Brien et al., 2003). The corresponding approach by the European Commission was based on measurements and resulted in the software EPCARD, which can also use the FLUKA simulation code

(Monte Carlo based) to interpolate experimental data (Schraube et al., 2002). The FLUKA Monte Carlo code itself is based on microscopic models (Fasso et al., 2001). PCAIRE code allows for dose-equivalent rate prediction at any geomagnetic latitude and altitude up to 20 km (Lewis et al., 2005). The model NASA AIR is based on a combination of theoretical calculations and experimental verification measurements (Wilson, 1991). The French government agency responsible for radiation protection of crew members has developed the software SIEVERT.

Although results obtained by different codes differ to some extent, there is reasonably good agreement between the results of calculations and experimental determinations (typically within 25% or better). In Figure 4.11, measured ambient dose equivalent is compared with calculated dose equivalent and effective dose equivalent for a trans-Atlantic flight (O'Brien et al., 2003).

The following codes can be downloaded from the Internet for dose calculations:

CARI	http://jag.cami.jccbi.gov/cariprofile.asp
EPCARD	http://www.gsf.de/epcard2/
PCAIRE	http://www.pcaire.com
SIEVERT	http://www.sievert-system.org/

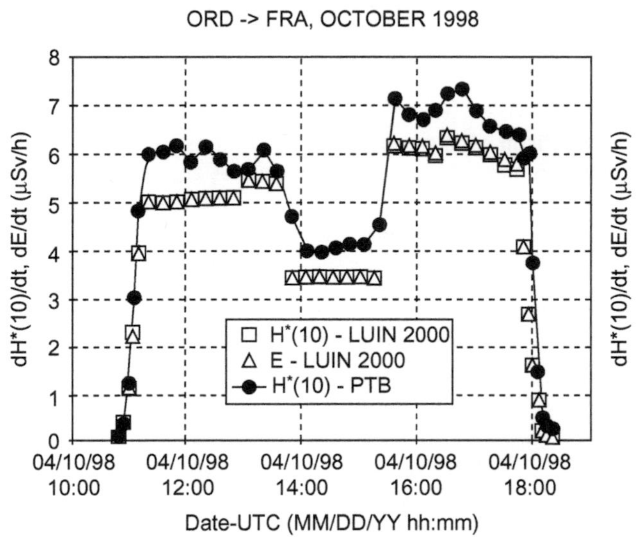

Figure 4.11 Comparison of measured ambient dose equivalent with calculated dose equivalent and effective dose equivalent for a flight from Chicago-O'Hare International Airport (United States) to Frankfurt am Main (Germany), in October 1998 (O'Brien et al., 2003). The authors are grateful to Keran O'Brien and colleagues for the permission to use this illustrative graph.

In order to improve dose assessments, microdosimetric quantities have also been determined for jet flights: frequency-mean lineal energy (y_F average lineal energy per event), dose-mean lineal energy (y_D), and the average quality factor (Q_F) (Lewis et al., 2001). In Table 4.3, these quantities are compared for ground-based sources and on commercial flights. For y_F the in-flight derived data correspond basically with ground-based data obtained from a ^{137}Cs source and the CERF exposure conditions. The in-flight Q-value of about 2 and the y_D value of about 14 indicate that in all flights there is a significant high LET contribution to the cosmic radiation field on board a jet aircraft flying at altitudes higher than 8.5 km. This is in contrast to occupational exposure conditions on the ground, which are largely determined by a Q-value of about 1. An in-flight Q-value of 2 was also found on long-haul flights of Belgian airlines (Verhaegen and Poffijn, 2000).

Exposure conditions of a crew onboard a jet aircraft are significantly different from the typical occupational radiation exposure in facilities on the ground: (1) Particle types and energies are different to typical occupational ground exposure. The high LET component of the absorbed dose of flight crews is about 50% as compared to a few percent for ground-based occupational exposure, as seen in Table 4.4. (2) Frequently female crew members represent the majority of crew members (or are represented, at best in equal numbers as compared to males), while they represent only a few percent of ground-based radiation workers. (3) Crew members are usually considerably younger than radiation workers on the ground. (4) It is possible to predict with a high degree of probability doses on board of jet aircraft (except for high-energy SPEs), while unplanned radiation exposure

Table 4.3. Comparison of microdosimetric quantities derived from ground-based and in-flight measurements (adapted from Lewis et al., 2001).

Source	Type of radiation	y_F (keV μm^{-1})	y_D (keV μm^{-1})	Q
Ground-based sources				
^{137}Cs	Gamma rays	0.36 ± 0.13	7.2 ± 0.8	1.1 ± 0.2
EU high-energy reference field (CERF)	Neutrons, muons, gamma rays	0.5	30 ± 3	3.8 ± 0.7
$^{239}Pu–^{9}Be$	Neutrons, gamma rays	1.39	51 ± 6	9 ± 2
^{244}Cm	Alpha particles	14.0	150 ± 20	24 ± 4
In-flight measurements above 8.5 km				
Trans-Atlantic	GCR plus secondary particles	0.358	14 ± 2	2.3 ± 0.4
Trans-Pacific		0.334	15 ± 2	2.2 ± 0.4
Trans-Canada		0.359	15 ± 2	2.3 ± 0.4

Table 4.4. Relative contribution of low and high LET components to the total dose equivalent on board a subsonic jet aircraft (Spurny et al., 2001).

Destination from Prague	Ratio of values high to low LET component[a]	Contribution of high LET component to the total[a]
North America	1.15	0.535
Northern Europe	1.09	0.521
Southern Europe	0.86	0.462
Abu Dhabi	0.514	0.338
Abu Dhabi–Bangkok	0.439	0.291

[a]Relative uncertainty of values: about $\pm 15\%$.

Table 4.5. Influence of solar activities on the total exposure level ($H^*(10)$) on board an Airbus 310-300 on the same flight route (adapted from Spurny and Daschev, 2002; Spurny et al., 2004).

Flight	Date	Event	$H^*(10)$
Prague–New York (JFK)	April 12, 2001	Forbush decrease	0.84
Prague–New York (JFK)	April 14, 2001	"Quiet"	0.98
Prague–New York (JFK)	April 15, 2001	Solar flare GLE 60	1.44

in ground-based occupational operations among radiation workers cannot be excluded even under normal operating conditions.

The relative importance of both components of the radiation field, low LET and high LET, on board a subsonic commercial jet is about equal in the Northern Hemisphere. The contribution by the high LET component becomes less important on routes approaching the Equator and reaches & ultimately about 30% (Table 4.4).

Another important parameter determining the radiation exposure in a commercial jet is the influence of solar activity extremes. In Table 4.5, the exposure levels on the same route are shown for a period of three days: (a) a "quiet" period as reference value, and for comparison; (b) during the solar flare GLE 60; and (c) a Forbush decrease. During the solar flare, the total exposure was about 40% higher than the reference value, while during the Forbush decrease, the corresponding value was approximately 26% lower.

One can notice in Figure 4.12 a factor of 2 increase in dose from $K_p = 1$ to $K_p = 4$. A pronounced effect on the effective dose rate results from the influence of the geomagnetic latitude; variations can reach one order of magnitude, while the variation due to solar activity is only about a factor of 2, as can be seen in Figure 4.13.

This latitude effect is clearly reflected in the relative contribution of the various particle types to the effective dose rate at different geomagnetic

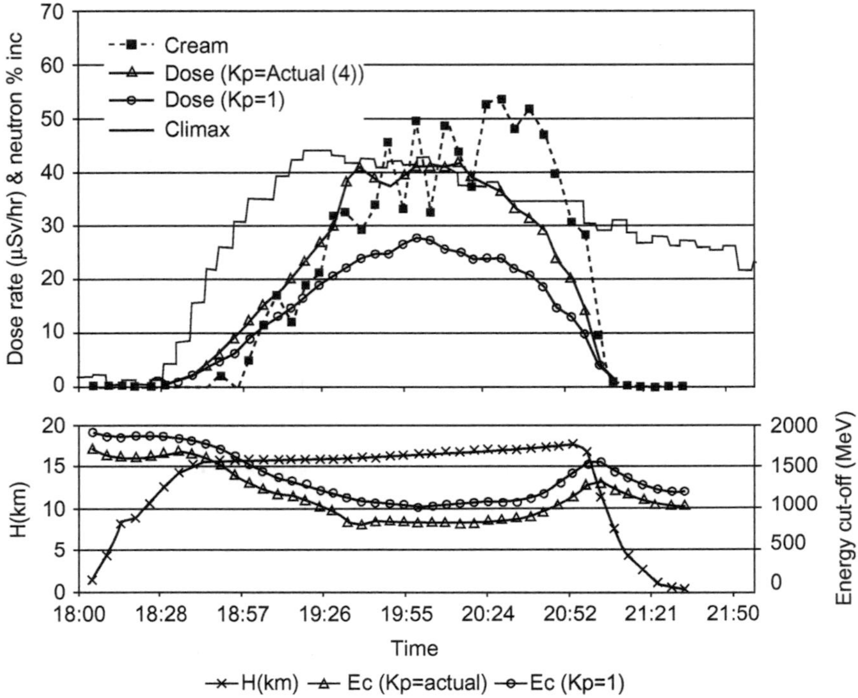

Figure 4.12 Dose rates from CREAM (cosmic radiation environment and activation monitor) on a Concorde flight from JFK (John F. Kennedy airport in New York) to LHR (London Heathrow airport) during the event of September 29, 1989 compared with calculations and climax neutron monitor rates (Dyer et al., 2003).

locations, as illustrated in Figures 4.13 and 4.14, each with its specific radiation-weighting factor. Due to the comparatively large w_R values of neutrons and protons, the effective dose is basically determined by these species.

Since the neutron component of the cosmic radiation represents at low latitudes about 50% of the dose equivalent at cruising altitudes of current commercial jets, information about the neutron spectrum is important (this increases to approximately 75% near the poles). Based on extensive measurements it could be shown that at high altitude, the effect of geomagnetic latitude on the neutron spectrum is small but rather large on the neutron fluence rate, that is, 7.8 times higher at 0.8 GV cutoff than at 11.6 GV. The change of the neutron spectrum with altitude from 12 to 20 km (atmospheric depth: 201–56 $g\,cm^{-2}$) is minor, but very different at ground level, which reaches over 1,000 $g\,cm^{-2}$ (Goldhagen et al., 2004a, 2004b).

This has significant impact on dosimetric calculations, determining effective dose and ambient dose equivalent ($H^*(10)$) rate, as well as

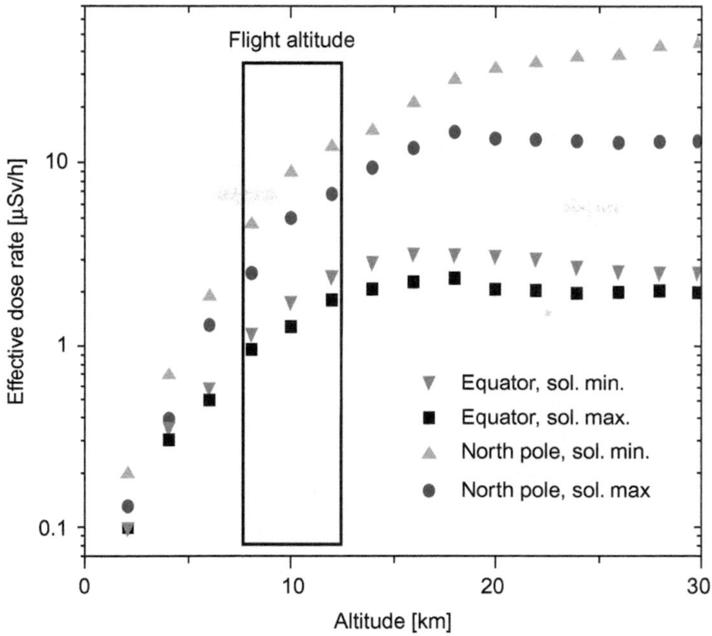

Figure 4.13 Effective dose rate as a function of altitude for two geomagnetic locations in different phases of the solar cycle (Ashkenazi et al., 2005) (reproduced by special permission of Dr. Ran Ashkenazi).

spectrum-averaged fluence-to-effective dose and fluence-to-H*(10) conversion factors. Table 4.6 contains the neutron integral quantities measured at different geographic locations and altitudes (Goldhagen et al., 2004a, 2004b).

At all locations and altitudes, ambient dose equivalent was found to be larger than effective dose for the neutron component in cosmic radiation.

The exposure pattern of cosmic radiation on aircraft personnel is further modified by the aircraft type they are working in. Secondary particles are produced by the breakup of cosmic ions and atoms of interstellar and atmospheric gas. A similar process occurs in aircraft skin. In order to account for the influence of aircraft shielding on the aircrew exposure inside, an aircraft model has been developed, based on an Airbus-340 (Ferrari et al., 2004). The model represents 162 regions with 17 different materials, describing fuselage structure, fuel, passengers, engines, and cargo. Table 4.7 shows the effective dose and ambient dose-equivalent rates for several locations inside the aircraft.

The effective doses received by crew members and passengers traveling on the same plane are dependent on location of the individual onboard, quantity of fuel loaded, and number of passengers onboard. The main results of the modeling show that (a) pilot and copilot are likely to receive a

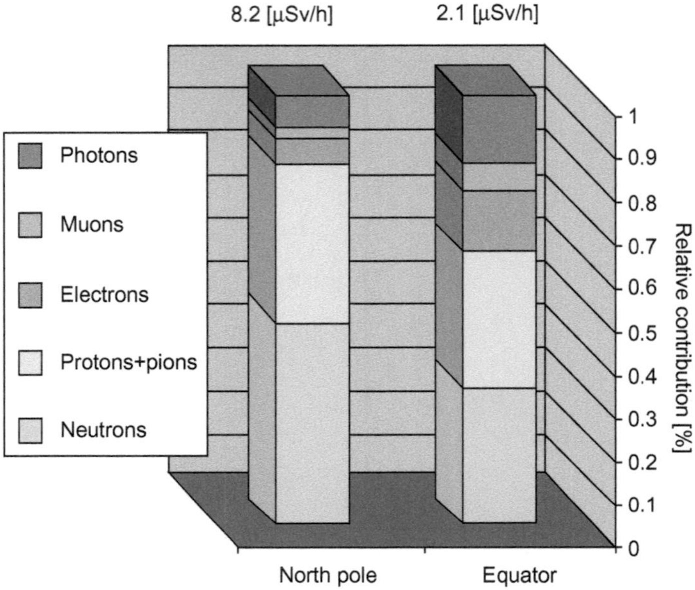

Figure 4.14 Relative contribution of different particles of the cosmic radiation to the effective dose rate at the North Pole and Equator (Ashkenazi et al., 2005) (reproduced by special permission of Dr. Ran Ashkenazi).

Table 4.6. Neutron integral quantities measured at different geographic locations and altitudes (Goldhagen et al., 2004a, 2004b).

Geographic location	Altitude (km)	Fluence rate ($cm^{-2}\,s^{-1}$)	Effective dose rate ($\mu Sv\,h^{-1}$)	$H^*(10)$ rate ($\mu Sv\,h^{-1}$)
54°N, 117°W	20.0	9.7	5.8	7.8
38°N, 122°W	11.9	3.4	2.05	2.7
37°N, 76°W	0	0.013	0.009	0.010

higher dose than flight attendants and (b) passengers in the middle of the plane receive a lower dose than those in window seats

An important parameter modifying the exposure pattern onboard an aircraft is the aircraft type (Tveten et al., 2000). An aircraft flies at a certain altitude, it is used on certain routes, and operates either on short-, intermediate-, or long-haul flights, that is, the influence of altitude and latitude reflect on the dose rate on board the aircraft. Table 4.8 shows the CARI-calculated dose rates for different types of aircraft. Independent of the period of operation of the aircraft, dose rates on board short-haul flights are lower ($\leq 0.5\,\mu Sv\,h^{-1}$) than on long-haul flights (1.7–$4\,\mu Sv\,h^{-1}$); intermediate-haul flights are somewhat less distinct from either category.

Table 4.7. Ambient dose-equivalent rates at the equator for several locations inside an Airbus-340 (altitude: 10,580 m, $R_v = 17.6$ GV, $\Phi = 465$ MV; adapted from Ferrari et al., 2004).

Location inside plane	$H^*(10)$ in $\mu Sv\,h^{-1}$
Pilot	1.22
Passenger in row 13, internal	1.02
Passenger in row 37, window	1.24

Table 4.8. Dose rates on board different aircraft as calculated by CARI-5E program (adapted from Tveten et al., 2000).

Year	Short-haul flights	Intermediate-haul flights	Long-haul flights
		$\mu Sv\,h^{-1}$	
1990	Fokker 50 (0.3)	MD-80 (2.0)	Boeing 767 (3.1)
1985	Fokker 27 (0.3)	Boeing 737 (2.3)	Boeing 747 (4.0)
1980	Twin Otter (0.08)	Airbus 300 (2.8)	DC 10 (3.5)
1975	Electra (0.5)	DC 9 (1.4)	DC 8 (2.5)
1970	DC 6 (0.2)	Caravelle (1.2)	DC 8 (2.8)
1965	Metropolitan (0.3)	Curtiss (0.5)	Coronado (1.7)
Range	0.08–0.5	0.5–2.8	1.7–4.0

Examples of typical doses received on various flight routes, together with the mean annual doses received by pilots and cabin crew, are summarized in Table 4.9. Depending on the duration and route of the flight, round-trip doses received by passengers and crew range from less than 1 μSv, for short national flights, to 40 μSv, for intercontinental flights between Europe and the US-West Coast. This translates to mean annual doses for the crew members of <1 mSv on short national flights, and up to >2 mSv for crew members working on trans-Atlantic routes between Europe and the United States.

In Figure 4.15, the cumulative doses of British Airways staff are shown for the period 1999–2001 as calculated using CARI-6 software (Irvine and Flower, 2005). The maximum average dose amounts to 4.25 $mSv\,y^{-1}$, experienced by a cabin crew member on the flight route Tokyo–London.

A special case is *high-altitude flights* (about 18 km) where crew and passengers receive elevated exposure to cosmic radiation, for example, on scientific balloon flights or onboard commercial supersonic aircraft, such as the British–French *Concorde* (Irvine and Flower, 2005).[2] Typically

[2] Altogether 14 such planes were operated by Air France and British Airways during the period from January 21, 1976 until April 10, 2003.

Table 4.9. Typical doses received on various flights and occupational annual doses of crew members.

Round-trip flights from/to	Effective dose per flight (μSv)	Mean annual effective dose for crew (mSv)	Period	Reference
Austria	USA (Atlanta): 4.1 Australia (Sydney): 2.1		2001	Hajek et al. (2005)
Lithuania[a]	0.5–13.4		2000–2001	Morkunasi et al. (2003)
Spain		North America: 2.22 Central America: 1.61 South America: 1.43 Africa: 1.25 Europe: 1.72 Spain[b]: 0.70	2001	Saez Vergara et al. (2005)
New Zealand	Southern Hemisphere: 8.3 Trans-equatorial: 9.3 Northern Hemisphere: 35.3		2002–2003	Taylor, et al. (2004)

Germany	USA Fairbanks: 40.0 Chicago: 38.5 New York: 30.9 UAE Bahrain: 5.0 Senegal Dakar: 13.9		1997–1999	Schrewe (2000)
USA		Chicago–New York: 4 Seattle–Portland: 0.2 New York–Tokyo: 7 Athens–New York: 9.1		Friedberg et al. (1989); Friedberg et al. (1993)
France		Paris–Tokyo: 5.9 Buenos Aires–Paris: 3.7		Bottolier-Depois (1997)

[a]European flights only.
[b]Excluding Canary Islands – Canary Islands 1.17 mSv.

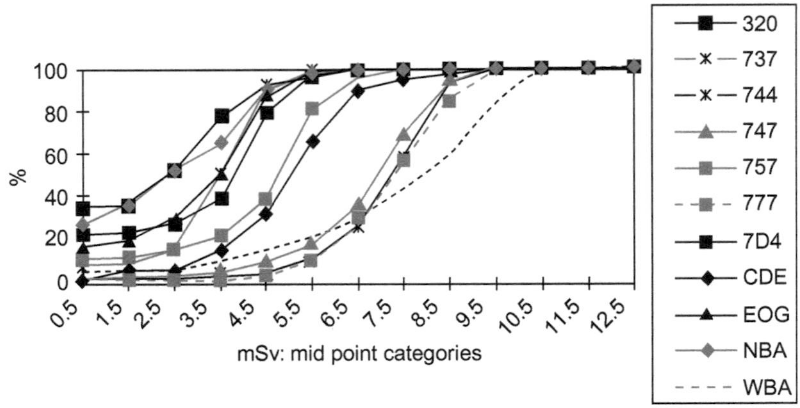

Figure 4.15 Cumulative doses of British Airways flight crew and cabin crew for different aircraft 1999–2001 (Irvine and Flower, 2005).

passengers and crew members on board such a plane receive an effective dose rate between 10 and 12 μSv h^{-1} (UNSCEAR, 1993). However, generally shorter flight times in supersonic planes result in lower total radiation doses per flight as compared to flight crews on subsonic planes operating on the same route.

The high-LET dose-equivalent portion amounts to about 41.5% (±8) with a mean Q_F of 1.65. Normally the dose contribution from SPEs is relatively small, for example, during the period December 1988 to July 1992 it amounted only to a 2 to 7% contribution of the total cosmic ray equivalent dose at 18 km altitude. It is possible that in the case of a highly energetic event, such as in February 1956, the dose-equivalent rate at 20 km can reach about 1 mSv h^{-1} (Armstrong et al., 1969). Under normal conditions, the ambient dose equivalent on a trans-Atlantic flight in a supersonic plane between Europe and the US East Coast ranges from 38 to 43 μSv (three-month moving average: period 1997–2000). Table 4.10 contains an overview of dosimetric data collected during the operation of *Concorde* (Bottollier-Depois et al., 2003).

4.3.2. Radiation exposure of space travelers

Currently it is almost exclusively astronauts on mission from the United States, Russia, Europe, and China who are exposed to cosmic radiation at higher altitudes (e.g., International Space Station (ISS), MIR Space Station), or during space travel to the Moon (US-Apollo missions). In the future, it is likely that due to space tourism a growing number of private space travelers will also be spending time at such altitudes. Furthermore, long missions are planned into deep space (e.g., Mars). The implications of increased exposure of space travelers to cosmic radiation are discussed below.

Table 4.10. Overview of dosimetric data collected during the operation of supersonic commercial flights on board the *Concorde* jet liner (Bottollier-Depois et al., 2003)[a].

Date/period	Route	Dose-equivalent rate ($\mu Sv\,h^{-1}$)	Annual mean dose equivalent (mSv)[a]	Comments
1976–1990	Paris–New York	10.6	Flight deck crew: 3.3, cabin crew: 2.7	Cruising height: 15,000–18,000 m[b]
August 21, 1996	Paris–New York	9.7 ± 1.5		
August 22, 1996	Paris–New York	9.2 ± 1.3		
1991–1992	Paris–New York		$2.6\ mSv \pm 0.5$	Minimum solar activity
1996–1998	Paris–New York		0.8 ± 0.5	Maximum solar activity
1988–1990	London–New York		3–6	
	All routes[c]	10–12		

[a]*Concorde* crew members operate only $300\,h\,y^{-1}$ instead of $700\,h\,y^{-1}$ for the flight crews working on subsonic aircraft.
[b]See Montagne et al. (1993).
[c]See UNSCEAR (2000).

4.3.2.1. Low-earth orbits

Astronauts working at the ISS orbit below the inner Van Allen belt closest to the Earth, except for the SAA[3]; this part of the inner radiation belt dips down lower to Earth than the rest of the radiation belt. The level of radiation depends on altitude and orbital inclination. In the case of the orbital inclination of the ISS (51.6°), the contribution of the GCR exceeds the equivalent dose resulting from protons trapped in the Van Allen belt. However, at the orbital inclination of many of the US space shuttle missions (28.5°), the trapped protons provide the larger component (Friedberg et al., 2005a, 2005b). Electrons in the outer radiation belt become more important during highly relativistic events. The electron energy is low and the radiation is only of concern to astronauts outside the spacecraft engaged in extravehicular activities at the higher orbital inclination, for example, during the construction of the ISS (Fry, 2002).

[3] The SAA, a region off the Brazilian coast, extends from about 0° to 60°W and 20° to 50°S (geographic coordinates) – see also Section 4.2.

4.3.2.2. Deep space missions

During missions to the Moon and interplanetary missions, astronauts are exposed to (1) charged particles which are trapped in the magnetic field of the Earth, (2) GCR (mostly protons and a lesser amount of heavy ions), and (3) SPE. While astronauts are residents on the Moon or on another planet, they will receive a radiation dose from GCR exposure. Astronauts receive from supernovae, as the main source of GCR, particles with energies $\geq 10^{20}$ eV (Bird et al., 1995). Also, in deep space astronauts may have to face a significant dose rate increase due to protons with high energy as a result of an SPE.

Tables 4.11 and 4.12 contain an overview of the exposure conditions as measured on board spacecraft in orbits around the Earth and, for comparison, as for a 2.7-year-long mission to Mars. The latter was

Table 4.11. Dose rates on board spacecraft in low-Earth orbits (Deme et al., 2005).

Location	Dose rate ($\mu Gy\,h^{-1}$)	Comments
On board spacecraft Salyut-6		5-day mission in 1980
On spacesuit of astronaut	3.5	
Passage	3.7–4.5	
Working area	2.8–3.8	
Sleeping place	4.3–4.5	
On board spacecraft Salyut-7		3 missions in 1983 (total duration: about 3 months)
Personal, working area, passage, sleeping place	6.2–7.5	Average of 15 dosimeters
On board MIR		Average over 5 missions in 1987 (total duration: about 28 weeks)
Cabin of commander	12.4 ± 2.0	
Cabin of board engineer	10.6 ± 2.2	
Passage	12.0 ± 1.8	
EVA	144–196	June 12 and 16, 1987
On board Space Shuttle		Mission no. 41 G at inclination 57° (duration: 197 h)
Altitude 270 km	3.5	
Altitude 270 km	6.0	
Altitude 350 km	10.0	
On board ISS		Second quarter of 2001
Average over several locations	5.6–8.3	

Table 4.12. Simulated 2.7-year duration mission to Mars (September 18, 2007–May 20, 2010).

Phase of mission	Time spent (days)	Effective dose (mSv)	Particle
Earth to Mars	280	880	GCR[a]
		170	SEP[b]
On Mars	439	410	GCR
Mars to Earth	256	800	GCR
Total	975	2.26×10^3	

Source: Data from Friedberg et al. (2005a, 2005b).
[a]Galactic cosmic radiation.
[b]Solar energetic particles.

modeled under the assumption of (a) avoiding anticipated times of highest solar activity and (b) launch times representing dates when Earth and Mars are positioned for the shortest travel time.

4.4. Effects of Cosmic Radiation on Avionics

When a primary particle hits a small volume of a target material, it can knock out electrons from atoms in its path due to its high energy (i.e., generate an electric charge). The rate of charge deposited per unit path length is proportional to the square of the atomic number of the cosmic ray particle. Thereby a heavier particle is able to deposit a sufficiently large charge in a silicon chip resulting in a change of the state of a memory cell from 0 to 1 and *vice versa*. Such corrupted memories can lead to errors in the execution of software commands (soft errors, single event upsets (SEU)) (Dyer and Truscott, 1999). If a single particle is able to change more than one bit, this results in a multiple bit upset (MBU). Several such single event effects (SEE) can lead to malfunctioning of electronic equipment: (1) triggering of a device into a state of high current drain, causing burnout and failure of the hardware affected (single event burnout (SEB)); (2) localized dielectric breakdown and rupture (single event breakdown (SED), single event dielectric failure (SEDF)).

Energy deposited per unit path length of material is defined as linear energy transfer (LET); in the case of radiation effects in electronics, the common unit is $\text{MeV}\,\text{g}^{-1}\,\text{cm}^{-2}$. For each electronic device, a threshold LET exists below which an SEE does not occur. With the continuing miniaturization of devices, their overall size shrinks and these thresholds move to lower LET.

Table 4.13. Damaging effects observed in electronic components and devices exposed to cosmic radiation.

Device	Component	Damaging effect	Comments	Reference
NASA/DoD Tracking and Data Relay Satellite (TDRS-1)	RAM chips	Upset rate: 1/day (quiet time); several hundred/day (solar particle events)		Wilkinson et al. (1991)
European Remote Sensing Spacecraft (ERS-1)	PRARE Hardware (Precision Ranging Experiment)	Loss of instrument after 5 days		Adams et al. (1992)
MIR	Unhardened IBM ThinkPad® computers	Upset rate: 1/9 h		Martignano and Harboe-Sorensen (1995)
Space Shuttle	Laptop computer	Upset rate: 1/hour		Klausman (1995)
Random Access Memory	A6U-15LL CMOS SRAM[a]	Bit errors accumulating at rate: 4.8×10^{-8}/bit, day	Exposed at 35,000 ft	Olsen et al. (1993)
Random Access Memory	IMS1601 SRAM	Bit errors accumulating at rate: 1.2×10^{-7}/bit, day	Exposed at 30,000 ft, 40° latitude	Taber and Normand (1996)
Random Access Memory	IMS1601 SRAM	Bit errors accumulating at rate: 5.4×10^{-7}/bit, day	Exposed at 65,000 ft, high latitudes	Taber and Normand (1996)
Random Access Memory	1 GB DRAM[b]	Upset rate: 1/3–1/14 h	Theoretical modeling	Kerness and Taber (1997)

[a]Complimentary metal oxide silicon static RAM.
[b]Dynamic RAM.

At cruising altitudes of ordinary commercial jet airliners, most PCRs are negligible and therefore induction of SEE by this mechanism has only a low probability of occurring. In the inner radiation belt of the Earth (with its intense flux of high-energy protons) and in the atmosphere (with its build-up of secondary neutrons), nuclear interactions can occur in addition to the directly ionizing interactions described above. Such interactions of particles with atomic nuclei will result in the transfer of recoil energy and the generation of secondary particles. Since both of these reactions are highly ionizing, they may result in an SEE in a device depletion region. Since 1975, when electronics became small and sensitive enough, the above effects have caused bit errors, unexpected error rates, non-repeatable errors, and computer crashes and failure of hardware in the space industry and even in systems installed at the surface of the Earth. In Table 4.13, the different effects observed in electronic components and devices are summarized. The damage ranges from bit errors accumulating at a rate of 4.8×10^{-8}/bit/day to the complete loss of an instrument installed on a satellite within less than a week of exposure to cosmic radiation.

CHAPTER 5

Metrology and Modeling

5.1. Metrology

The International Bureau of Weights and Measures (in French – Bureau International de Poids et Measures – BIPM) defines metrology as "the science of measurement, embracing both experimental and theoretical determinations at any level of uncertainty in any field of science and technology" (BIPM, 2005). As is well put by the BIPM, wrong and inaccurate measures can lead to wrong decisions. The consequences of wrong decisions can be enormous, including large financial losses, economic disadvantage, or even human death and suffering. Metrology is a very old science; however, the distinction between legal and scientific metrology is to be carefully distinguished today. This chapter will deal only with scientific metrology as far as measurements of NR are concerned.

Measurements of any characteristic of the electromagnetic spectrum, like wavelength, energy, or frequency, pose remarkable difficulties because there are inherent uncertainties associated with most of these characteristics.

5.1.1. Prehistory of the metrology of ionizing radiation

The international measurement standards for ionizing radiation were first adopted by the BIPM in 1960 (see BIPM, 2005). However, the history of the metrology of ionizing radiation may be considered to have started soon after the discovery of radioactivity.

The historic (or prehistoric) account of the metrology of ionizing radiation presented in this section is based on a paper presented by one of the authors (ASP) at the Metrologia 2000 held in São Paulo, Brazil (Paschoa, 2000a, 2000b). It must be remembered that natural radioactivity was present in all of the experiments with ionizing radiation carried out just after the discovery of radioactivity.

As is well known, by the end of the year 1895, Röentgen discovered X-rays in Wurzburg, Germany (Glasser, 1934; Dibner, 1968). Less than 1 month after, on January 20, 1986, Röentgen wrote to Poincaré to ask him to report that discovery at a meeting of the French Academy of Sciences in Paris (Cassette, 1996). Becquerel, who was present at that meeting, went to

Radioactivity in the Environment, Volume 17
ISSN 1569-4860, DOI 10.1016/S1569-4860(09)01705-7

his laboratory to test the new discovery in a number of ways (Cassette, 1996; Paschoa, 1997). Among the various substances that Becquerel used to test Röentgen's discovery was a phosphorescent salt, a double sulfate of uranium and potassium, placed on a photographic plate wrapped in a thick, black cardboard sheet exposed to the sunlight. Becquerel expected that the sunlight would produce luminescence, a phenomenon that had been a subject of investigation by his family for two generations (Cassette, 1996; Paschoa, 1997). The uranium salt marked the photographic plate, inducing Becquerel to think that phosphorescence and X-rays were somewhat connected (Cassette, 1996; Paschoa, 1997). However, after performing some other experiments, Becquerel reported to the French Academy of Sciences that uranium crystals in contact with photographic plates, but kept in darkness, would produce the same impressions as the X-rays (Becquerel, 1896). Later on, Marie Curie repeated Becquerel's experiments with radium and termed the phenomenon radioactivity. The communication of the discovery of a new element called radium, and the introduction of the term radioactivity was made by the Curies and Bémont in a session of the French Academy of Sciences on December 26, 1898 (Curie et al., 1898), and was described by Eve Curie (Curie, 1937) as follows:

> The various reasons we have just enumerated lead us to believe that the new radioactive substance contains a new element to which we propose to give the name **RADIUM**.
> The new radioactive substance certainly contains a very strong proportion of barium; in spite of that, its radioactivity is considerable. The radioactivity of radium therefore must be enormous.

The explanation for the phenomenon of radioactive disintegration came just a few years later through the work of Ernest Rutherford and Frederick Soddy (Rutherford and Soddy, 1902), who proposed that elements like radium transmuted themselves into other elements, yielding energy in the process. As a matter of fact, Rutherford explained the decay chains of the three main natural series as soon as they were identified (Rutherford, 1904, 1905).

Even before Rutherford identified a radioactive gas called "emanation" (Rutherford, 1900), today known as radon (^{222}Rn), Debierne (1899) had discovered actinium (^{227}Ac), and 4 years later, found other emanations (Debierne, 1903), now known as the radioactive gases thoron (^{220}Rn) and actinon (^{219}Rn).

Here it is worth mentioning that those scientists who discovered and carried through the earlier works to characterize the naturally occurring radionuclides were the true initiators of what is known today as the metrology of ionizing radiation. The alpha particles emitted by the heavier radionuclides and the beta particles were studied by Rutherford and other investigators to determine their relation E/m, where E is the electric charge and m the mass, and their velocity v.

A typical device used by Rutherford to characterize alpha particles is represented in Figure 5.1. The first and second tracks C_1 and C_2 that appear in the figure are produced by the end of the trajectories of alpha particles submitted to a magnetic field, and then inverting it. Using a device like this, Rutherford obtained $E/m = 5 \times 10^3$ charge units per gram, and $v = 2.0 \times 10^7\ \mathrm{m\,s^{-1}}$.

Regener (1909) built a counting system for the alpha particles emitted by radium F (^{210}Po) using zinc sulfide (ZnS) deposited on a backing (or alternatively on a 100 μm layer of diamond), as represented schematically in Figure 5.2. The ^{210}Po source was maintained in vacuum, and a scintillation caused by an alpha particle reaching the ZnS was observed and counted through an optical microscope, as shown schematically in the figure. This was, for sure, the first scintillation counter ever built. The counting uncertainties were not available.

5.1.2. Historical aspects of radon metrology

At the beginning of the 20th century, the amount of a radionuclide present in water was expressed by weight in grams, or in submultiples of 1 g. The exception was radon and its progeny. Irrespective of the chemical form of radium in water, one would say that a water sample would contain $0.1\ \mathrm{pg\,L^{-1}}$ (1×10^{-11} g) ^{226}Ra.

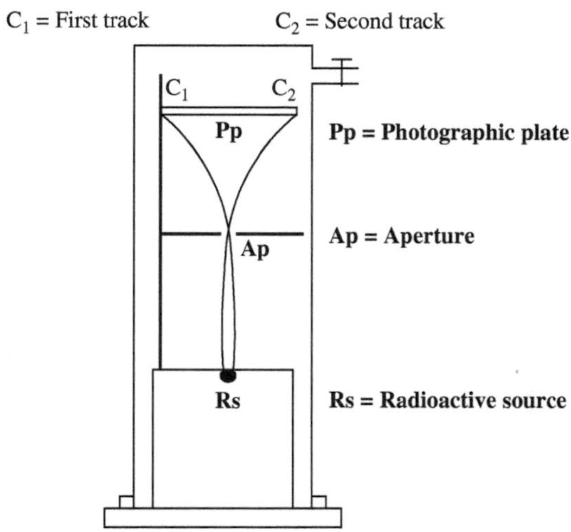

Figure 5.1 Rutherford-like device developed to measure E/m and v of an alpha particle. Adapted from Castelfranchi (1934).

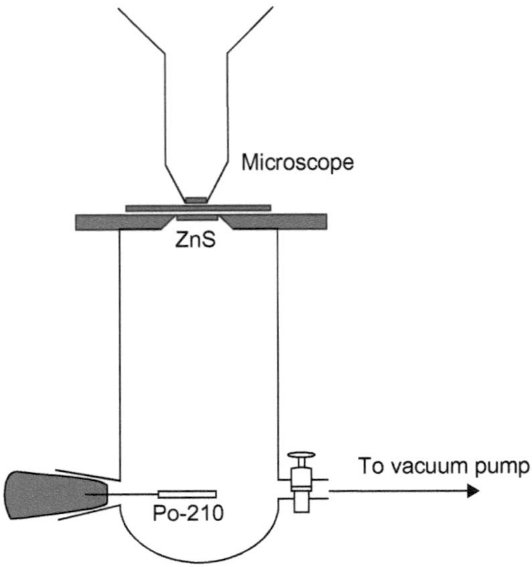

Figure 5.2 Regener-like device developed for counting alpha particles. Adapted from Castelfranchi (1934).

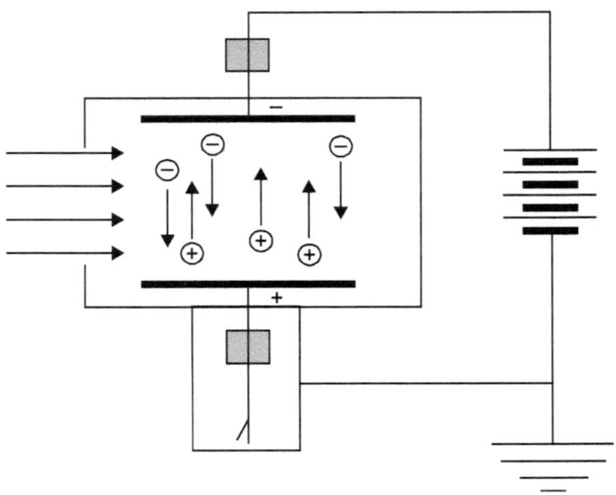

Figure 5.3 Schematic representation of an air chamber (or electrometer) of the type used to measure radon at the beginning of the 20th century. Adapted from Privault (1934).

In 1904, in Germany, the *mache* unit was introduced to measure radon in a fluid, based on the ionization current in an air chamber, of the type shown schematically in Figure 5.3. One *mache* is, by definition, the radon concentration in 1 L of a fluid, which corresponds to the presence of a quantity of radon able to produce by itself (i.e., excluding the disintegration products), and using completely the energy of the alpha particles emitted, a saturated ionization current equal to 0.001 of the electrostatic unit. The ionization current was measured by a galvanometer similar to that used by Curie (1908), illustrated in Figure 5.3, which was a popular device to measure radioactivity at the beginning of the 20th century (Laborde, 1911).

In 1910, however, at the 2nd International Congress of Radiology, held in Brussels, it was established that one unit of radioactivity of the gaseous ^{222}Rn would be a quantity of this nuclide in equilibrium with 1 g of ^{226}Ra, which meant 6.47 μg ^{222}Rn. This unit of radioactivity became known worldwide at a later date as 1 Curie (1 Ci) (see, e.g., Paschoa, 2000a, 2000b).

Taking into account that the weight of an element in a sample is proportional to the number of atoms of that element in that sample, plus the equations of radioactive equilibrium (Castelfranchi, 1934), one can conclude that 1 Ci ^{222}Rn contains 1.77×10^{16} atoms in a volume of 0.66 mm^3, at 0° C and 760 mm Hg, and, by definition of equilibrium, emits the same number of alpha particles per second as 1 g ^{226}Ra.

In May 1921, at the Congress of Radiology, held in Freiburg, Germany, a unit, called an *eman*, was introduced. This was considered convenient for measuring radon in hydrothermal fountains. In 1931, the International Committee for the Standard of Radium established finally that, based on Avogadro's number and the atomic number of ^{226}Ra, the atom-gram of ^{226}Ra contains 2.68×10^{21} atoms, which emits 3.7×10^{10} alpha particles per second. This number is, of course, used for the definition of 1 Ci.

Table 5.1 presents the equivalences among the units *mache*, *eman*, Curie, and Becquerel (Bq).

An interesting account of the evolution of internationally adopted radioactivity quantities, units, and standards was given by Allisy (1994/1995).

Table 5.1 Equivalence among the old and current units used for measurements of radon in water.

	mache	pCi L^{-1}	eman	Bq m^{-3}
1 mache	1	364	3.64	1.35×10^4
1 pCi L^{-1}	0.0028	1	0.01	37
1 eman	0.28	100	1	3.7×10^3
1 Bq m^{-3}	7.4×10^{-5}	0.027	2.7×10^{-4}	1

Source: Taken from Paschoa (2000a, 2000b).

5.1.3. Metrology of ionizing radiations before WWII

A number of devices to measure ionizing radiation were developed during the first half of the 20th century. Before the beginning of WWII, investigators thought that much was already known about the properties of the naturally occurring radionuclides. Conversely, very little was known about the fission products, although the phenomenon of fission had been discovered in 1934 (Fermi, 1934; Fermi, 1954; Hahn and Meitner, 1935; Hahn, 1958). Readers interested in details regarding what was known about fission at the beginning of WWII are directed to the excellent review published by Turner (1940).

The Rutherford and Geiger ionization chamber of 1908, the Geiger–Muller particle counter, among other devices based on the detection of ionizing radiation, was commonly used well before WWII. As early as 1903, medical doctors knew that X-rays could provoke biological effects, such as cancer, sterilization, and damage to the bone marrow. The accepted view, however, was that the benefit of using X-rays would exceed the risks.

It is well known that the ions produced in a gas may be accelerated by a bias voltage to produce X-rays. However, when X-rays were used at first, the gas pressure and the voltage would vary unpredictably, making the radiation also vary in an erratic way from exposure to exposure and from minute to minute. Trial and error was the only way to make progress with the use of X-rays at that time. The energies transmitted and absorbed, among other quantities, were related to the so-called dose. However, erythema (reddened skin) and epilation (hair loss) were the dose indices that deserved the confidence of the earlier X-rays workers. Later, the medical doctors tried to irradiate patients at shorter intervals to avoid reddened skin (Parker and Roesch, 1963).

In Germany, between 1914 and 1924, physicists and physicians started to question some aspects of the dosimetry of ionizing radiation, such as: What are the main sources of uncertainties in the measurements of ionization? Do biological effects depend essentially on the absorbed energy? It was then necessary to measure the level of ionization that would cause reddened skin after a certain irradiation time and to adopt such a level as a biological unit. The physicists objected that the physically produced ions should be measured in absolute physical units, not in biological units. The important questions of the 1920s were the following: What quantities of ionizing radiation can cause damage? What quantities of ionizing radiation can be tolerated?

Mutscheller (1925) suggested that a worker would be entirely safe if he would not receive every 30 days a dose exceeding 1/100 of the erythema dose. This is the origin of what became known as the *tolerance dose*. Later, in the International Congress of Radiology held in Stockholm, the roentgen (R) became the adopted international unit for measurements of

X-rays, based on the ionization of the air (Taylor, 1958). Thus, the roentgen was a unit of exposure, but not of radiation absorbed dose. However, by 1934, the International Commission on Radiological Protection (ICRP) proposed a dose rate limit of 0.2 R per day.

For the reader interested in the history of metrology, it is also recommended to read the article on the evolution of the standard for ^{226}Ra from Marie Curie to the International Committee for Radionuclide Metrology (Coursey et al., 2002).

The fact that some chemicals change color when irradiated with X-rays made the clinics adopt them as indicators of the amount of radiation received by patients. Such chemical indicators of dose were used for decades (Hacker, 1987). The physicists, however, preferred the use as an index of the ionization of air and other gases. The ion currents measured, by various instruments, the degree of ionization, the ion number, and the intensity of the X-rays (Failla, 1923). However, only during and after WWII did the ionization methods to measure radiation become universally accepted (Hewlett and Anderson, 1962).

Merril Eisenbud called it a "remarkable coincidence" that the standards for radiological protection appeared just before they became extremely needed because of the Manhattan Project (Eisenbud and Gesell, 1997). Eisenbud was referring to the adoption of the roentgen unit 7 years before the beginning of the Manhattan Project.

5.1.4. Metrology of ionizing radiation in the Manhattan Project

As became well known after the end of WWII, the code name of the mammoth effort that encompassed a number of secret activities during that war in the United States of America was the Manhattan Project. When the Manhattan Project was initiated in 1941, little was known about the properties of the naturally radioactive elements, although the phenomenon of radioactivity had been discovered for almost half a century. Most scientists, however, thought they knew a lot about these properties. In any case, much less was known about the properties of the radioactive materials used and/or produced in the Manhattan Project, such as: plutonium, enriched uranium, and fission products. The quantities of these materials to be handled were much higher than those of ^{226}Ra and ^{228}Ra ever used in the experiments carried out by the Curies, Rutherford, and many others. The atomic pile developed by Fermi in Chicago, which converted uranium into plutonium, also produced radioactivity at concentration levels much higher than any known thus far. Plutonium had been discovered by Seaborg (1958) at Berkeley, but very little was known about the properties of this element.

Radioactivity was then a threat to the health of hundreds of technicians and scientists at work in the Manhattan Project. Few of them, however, had any knowledge of the hazards to which they were exposed. In addition, little could be told to them because of the secrecy involved in the Project. The nuclear weapon producers had anticipated some of the problems and agreed earlier in the Project to protect the workers and the public against unnecessary exposures to ionizing radiation.

At the beginning of the Manhattan Project, the military planners, with knowledge of the use of X-rays to locate bullets in WWI, chose pre-WWII standards already adopted in Radiology to protect the workers. At the beginning of WWII, the most common methods available for the detection of alpha and beta particles, gamma and X-rays, and neutrons were the following: scintillation; electroscopes; ionization chambers; proportional counters; Geiger–Muller counters; Wilson chambers; photographic emulsions; and the nuclear reaction $^{10}B(n,\alpha)^{7}Li$ to measure neutrons. All these methods were used during WWII.

What the general public does not generally know, however, is that the modern metrology of ionizing radiation is in fact one of the spin-offs of the Manhattan Project. By far the main goal of this Project was to produce the first nuclear weapons. Some of the problems and solutions concerning the development of the modern metrology of ionizing radiation during, and immediately after, WWII are directly related to the Project.

The first established standards adopted for protection of workers at the beginning of the Manhattan Project were 0.1 R per day of X- or gamma rays, 0.1 μCi (3.7 kBq) ^{226}Ra, or 10 pCi $^{222}Rn\,L^{-1}$ (100 Bq $^{222}Rn\,m^{-3}$) of air. Uranium, for example, was not considered to be a problem from the radiological point of view. After all, the half-life of ^{238}U was measured in billions of years and that of ^{235}U in hundreds of millions of years. Thus, both uranium isotopes were not considered to be a large problem, even if they were ingested (Hacker, 1987). In 1940, the Victoreen Instrument Co. of Cleveland, OH, produced the first commercial miniaturized ionization chamber and this was widely used at the beginning of the Project for the protection of workers (Lapp, 1949). The other type of personal dosimeter was the photographic film, but the latter was used just to indicate the accumulated dose over weeks (Hacker, 1987). Two systems of films were developed, one sensitive to a range from 0.03 to 3 R and the other covering the range from 1 to 20 R (Hacker, 1987).

The concept of *tolerance dose* continued to be used well after WWII, although the expression *maximum permissible exposure* had started being used already in the mid-1930s. The conceptual difference was that *tolerance dose* implied a lack of biological effect while *maximum permissible exposure* to X- or gamma rays represented an upper limit that should not be exceeded. However, the dose limits were not defined either for alpha and beta radiation or for neutrons.

By July 1942, Arthur H. Compton had already decided to put together a team to take care of radiation protection at the Metallurgical Laboratory in Chicago. There, on December 2, 1942, the first fission chain reaction took place in the atomic pile built by Fermi (Hewlett and Anderson, 1962). Later on, other facilities of the Manhattan Project had teams involved in research on radiation protection.

The measured quantity *roentgen equivalent physical* (rep), meaning the energy absorbed per unit mass (ergs per gram) equivalent at a point in the body to exposure in roentgens, was introduced in WWII. A biological factor – *relative biological effectiveness* (RBE) – was also introduced, in 1944, to take into account the fact that some biological effects varied with the kind of rays. However, the concept of RBE remained in secrecy until 1948 (Hacker, 1987). The product of the rep times the RBE became known as the *roentgen equivalent mammal or man* (rem). By 1958, the rem was an internationally accepted unit.

The 1944 preparation plan for the invasion of Europe by the Allied forces, code named "Peppermint," included several radiation survey meters, a Geiger counter, and a large number of film packets (Hacker, 1987). General Dwight Eisenhower and his staff had been briefed on potential radiation hazards should the Germans use radioactive poison against the Allied forces in Normandy.

When the Trinity test was ready to start at the beginning of July 1945, everyone involved in the test carried pocket dosimeters. Pocket dosimeters and film badge readings recorded exposure rates as low as $0.1\,\mathrm{R\,h^{-1}}$ at about 1,500 meters south of ground zero. Nearer ground zero, the readings of survey meters went off-scale. In some cases, only estimates of the order of at least $600\,\mathrm{R\,h^{-1}}$ could be made (Hacker, 1987).

Today, all units of radioactivity measurement, to express doses, are in agreement with the Systéme Internationalle (SI). This, however, is another history.

5.1.5. Current NORM Metrology

The very fact that the Member States of the European Union as well as the United States have decided to regulate NORM/TENORM created a metrology problem (USEPA, 1994; EC, 1996a; EU, 1996). The US Environmental Protection Agency's Radiological Protection Guide (USEPA/RPG) proposed recommendation of a limit of 100 mrem (1 mSv) effective dose equivalent received by, or committed in a single year to, any individual from all sources combined, implying the need to determine NORM/TENORM just at the NR background levels. As a consequence, the absence of a well-defined baseline of exposure from natural sources of radiation against which exposure to NORM/TENORM (or TENR in general) can be compared makes the associated metrology very difficult to deal with. In the case of the United States, the additional recommendation that doses from individual sources should normally be limited to a fraction of the USEPA/RPG plus the need to emphasize the principle that all exposures be as low as reasonably achievable makes the metrology of natural radionuclides almost intractable. Fortunately for those dealing with NORM/TENORM measurements and calibration, a small collection of documents published in the 1990s and 2000s made clarifications concerning the NORM/TENORM issues: (i) principles laid down by Title VII of the Council Directive 96/29/EURATOM to deal with a significant increase in exposure due to NR sources (EURATOM, 1996; EC, 1996a); (ii) IAEA Technical Reports Series No. 49 (IAEA, 2003a); and (iii) IAEA Safety Reports Series No. 34 (IAEA, 2003b).

Table 5.2 presents a list of national ionizing radiation metrology laboratories in some countries. The International Organization for Standardization (ISO)

Table 5.2 List of selected national ionizing radiation metrology laboratories.

Country	Laboratory	Name
Australia	ARPANSA	Australian Radiation Protection and Nuclear Safety Agency
Brazil	INMETRO/ LNMRI	National Institute of Metrology Standardization and Industrial Quality/ National Laboratory of Ionizing Radiation Metrology
China	NIM	National Institute of Metrology
France	BNML-LNHB	Bureau National de Métrologie – Laboratoire National Henri Becquerel
Germany	PTB	Physikalisch Technische Bundesanstalt
Greece	HIRCL-GAEC	Ionizing Radiation Calibration Laboratory of the Greej Atomic Energy Commission
India	BARC	Bhabha Atomic Research Centre Standards Section
Italy	ENEA-INMRI	Ente per le Nuove Technologie, L'Energia e L'Ambiente – Instituto Nazionale di Metrologia delle Radiazioni Ionizzanti
Japan	AIST (NMIJ)	National Metrology Institute of Japan
Romania	IFIN-HH	"Horia Hulubei" Institute of Physics and Nuclear Engineering
South Africa	NMISA	National Metrology Institute of South Africa
UK	BSI	British Standard Institution
Ukraine	SC-7	Scientific Centre of Ionizing Radiation and Nuclear Constants
United States of America	NIST	National Institute of Standards and Technology

develops standards for ionizing radiation in cooperation with other international organizations like the BIPM and the IAEA.

Calibrations for radiation measuring instruments need to be traceable to national or international standards. Ideally, each country with radiation activities must have a national laboratory in which reference standards are kept and made available for calibration.

X- and gamma-ray emission probabilities per decay of ^{238}U, ^{235}U, and ^{232}Th in equilibrium (or not) with their respective progenies are quite important for NORM/TENORM/TENR measurements and calibration. A number of articles have addressed these relevant issues for NORM/TENORM/TENR (Zobel et al., 1977; Farouk and Al-Soraya, 1982; Olson, 1983; Schötzig and Debertin, 1983; Lin and Harbottle, 1991; Morel et al., 1998; Delgado et al., 2002). Moreover, the ^{226}Ra activity (~40 kBq) in a gamma-ray spectrometric source was measured in relation to an alpha particle spectrometric source with an uncertainty of only 0.5% (Kharitonov et al., 2002).

The standardization of ^{222}Rn is a difficult task. However, an interesting attempt was made at the standardization of ^{222}Rn by liquid scintillation (LS) counting and comparison with alpha and gamma spectrometry (Cassete et al., 2006). Comparison between measurements made by: (i) gamma spectrometry with a LS reference ^{222}Rn source with 8.881×10^2 Bq activity agreed within 2.4%; and (ii) alpha spectrometry with a LS reference source with 6.456×10^2 Bq activity agreed within 1.23% (Cassete et al., 2006).

In Brazil, for example, 24 laboratories participated in an intercomparison exercise with a view to accreditation for a number of radionuclides. Among the latter, the following natural radionuclides were analyzed: U natural; ^{238}U; ^{234}U; ^{234}Th; ^{226}Ra; ^{214}P; ^{214}Bi; ^{210}Pb plus ^{232}Th; ^{228}Ra; ^{228}Ac; ^{212}Pb; ^{212}Bi; as well as ^{40}K (Vianna et al., 1998; Tauhata et al., 2002, 2006). Figure 5.4 represents in the form of a graph the analytical performance (%) as a function of the dates of intercomparison obtained in Brazil from 1991 to 1995 involving 955 analyses (Vianna et al., 1998). The results were analyzed in accordance with the statistical criteria used by the USEPA (Jarvis and Siu, 1981). Natural radionuclides comprised about 30% of these analyses (i.e., 293 out of 955). Of the results for the 11 laboratories that participated in these 293 analyses of natural radionuclides, 73.8% were considered good, 10.7% were acceptable (within warning limits), and 15.5% non-acceptable (out-of-control limits). Figure 5.5 shows the number of participant laboratories, the number of radionuclides analyzed, and the number of matrices used per year in the Brazilian National Program of Intercomparison – "Programa Brasileiro de Intercomparação (PNI/IRD)" from 1991 until 2004. The USEPA provided samples for the PNI/IRD from 1991 to 1995, and in 1996, the Environment Measurement Laboratory of the United States Department of Energy (EML/DOE)

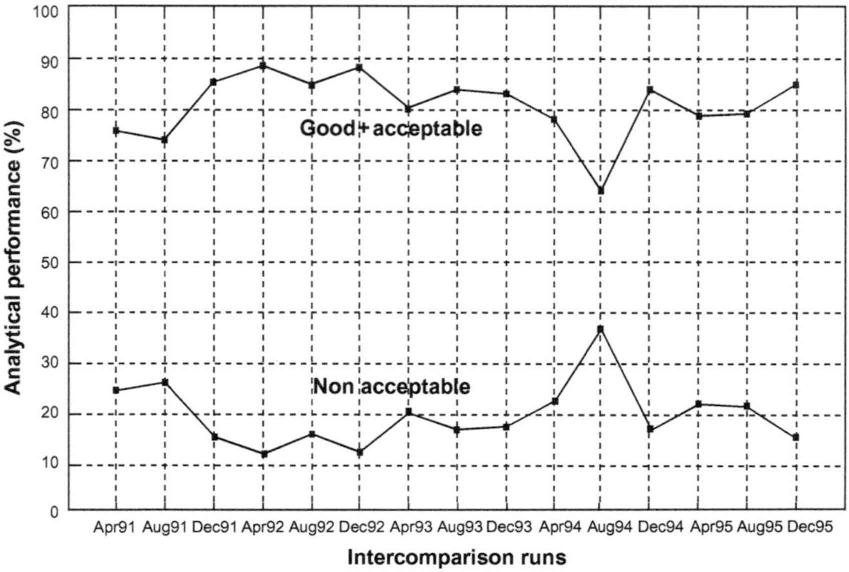

Figure 5.4 Graph of the analytical performance of 955 Brazilian laboratories as a function of the dates of the intercomparison between 1991 and 1995 (Vianna et al., 1998). The figure is reproduced here with the kind permission of Maria Elizabeth C. M. Vianna.

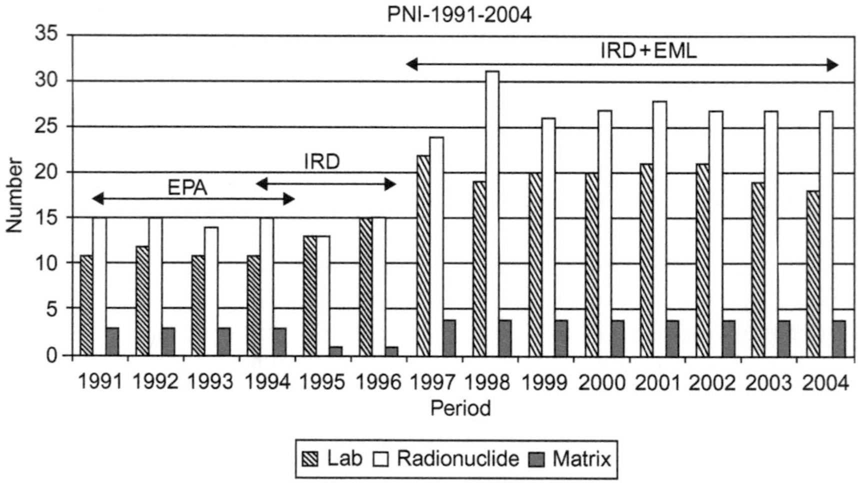

Figure 5.5 Graph of the number of participant laboratories, the number of radionuclides analyzed, and the number of matrices used per year in the Brazilian National Program of Intercomparison – "Programa Brasileiro de Intercomparação (PNI/IRD)" from 1991 until 2004. The figure is reproduced here with the kind permission of Luiz Tauhata.

started providing samples for the intercomparison exercise. The continuous analytical improvement of the participant laboratories allowed the good plus acceptable performance group of laboratories to reach over 90% in recent years.

Increasing public awareness of the NORM/TENORM/TENR problem pressured legislative bodies of countries and international organizations to adopt specific regulations to deal with the issue. It was anticipated that among the consequences and requirements of the NORM/TENORM/TENR regulations would be the following (Paschoa, 2001):

1. lowering of the legally tolerable concentrations of natural radionuclides in air, soil, water, and food;
2. need for measuring concentrations of natural radionuclides at levels barely above those considered to be normal in environmental and biological samples;
3. requirements for assessment of tolerable concentrations of radionuclides in environmental and biological samples after radiological emergencies and nuclear accidents;
4. requirements for assessment of chronic exposure to ionizing radiation either due to NR or as a consequence of orphan sources;
5. need for individual and team training to respond to increasing demands for assurance in radiation protection and radioecology.

The above list summarizes the challenges that NORM/TENORM/TENR metrology will face in the foreseeable future, as the regulations start being applied throughout the world. Of particular interest will be the application of NORM/TENORM/TENR regulations in developing countries with NORM extraction industries.

There will be two forthcoming international scientific and technological gatherings that will deal specifically with NORM/TENORM/TENR issues:

1. NORM and Natural Radiation Management: Strategies – Approaches – Techniques, to be held in London, May 28 and 29, 2009; and
2. 6th International Symposium on Naturally Occurring Radioactive Material (NORM VI), to be held in Marrakech, Morocco, March 22–26, 2010.

It is expected that sessions in the forthcoming NORM meetings will deal objectively with the issue of metrology, as well as other extant issues regarding NORM/TENORM/TENR.

5.2. Modeling

5.2.1. Environmental modeling as a tool

Environmental modeling may be helpful to deal with some problems related with NORM/TENORM/TENR. Table 5.3 summarizes the activity concentrations of ^{40}K, ^{232}Th, and ^{238}U in major rock types and soils. Data in Table 5.3 refer to unaltered rocks and soils of normal areas (i.e., data from high natural radioactive background areas are not included). The levels of radioactive concentration presented in Table 5.3 are not subject to any radiological control.

After mineral extraction and processing, activity concentrations may be considerably higher than those in unaltered rocks and soils. Table 5.4 presents a list of activity concentration levels found in selected mineral ores, which are sources for industrial exploitation.

The activity concentrations of NORM can be enhanced well above the natural background for a number of reasons. The activity concentrations presented in the last column of Table 5.4 illustrate the wide variety of values found in ores, products, and by-products of a diversity of industrial sources. In some cases, consumer products may have enhanced concentrations of NORM without the knowledge of the user. The release of TENORM into the biosphere is a current problem for scientists, engineers, industrial managers, and society in general to deal with.

In addition, physicochemical transformations may make NORM/TENORM more biologically available in the environment, enhancing its

Table 5.3 Activity concentrations of natural radionuclides in some rocks and soils.

	^{40}K	^{232}Th	^{238}U
	Bq kg^{-1}		
Igneous			
Basalt	3.0×10^2	10–15	7–10
Granitic	1.0×10^3	70	40
Sedimentary			
Shale, sandstones	8.0×10^2	50	40
Quartz	$<3.0 \times 10^2$–4.0×10^2	<8–25	<1–40
Arkose	6.0×10^2–9.0×10^2	<8	10–25
Beach sands	$<3 \times 10^2$	25	40
Carbonate rocks	70	8	25
Upper crust	8.5×10^2	44	36
Soils	4.0×10^2	37	66

Source: Adapted from IAEA (2003a).

Table 5.4 List of activity concentrations of selected mineral ores, products, and by-products.

Element or mineral	Source	Activity concentration
	Ore	2.5×10^2 Bq kg^{-1} (U)
	Bauxitic limestone (soil)	1.0×10^2–4.0×10^2 Bq kg^{-1} (Ra)
	Bauxitic limestone (soil)	30–1.3×10^2 Bq kg^{-1} (Th)
	Tailings	70–1.0×10^2 Bq kg^{-1} (Ra)
	Ore	30–1.0×10^5 Bq kg^{-1} (U)
	Ore	20–1.1×10^2 Bq kg^{-1} (Th)
Fluospar	Tailings	4.0×10^3 Bq kg^{-1} (Ra)
		6.0×10^3–2.0×10^4 Bq kg^{-1} (U)
		Th series – 4% by weight
	Gas – average – (USA and Canada)	2–1.7×10^4 Bq m^{-3} (Rn)
	Gas – individual well – (USA and Canada)	0.4–5.4×10^4 Bq m^{-3} (Rn)
	Scale – residues in pumps, vessels, and residual gas pipelines	1.0×10^2–5.0×10^4 Bq kg^{-1} (^{210}Pb, ^{210}Po)
	Brines or produced water	1.0–1.0×10^5 Bq m^{-3} (Ra)
	Sludges (scales)	Up to 7.0×10^4 Bq kg^{-1} (Ra)
	(Scales)	Typically 1.0×10^3–1.0×10^4 Bq kg^{-1} (Ra), and up to 4.0×10^6 Bq kg^{-1} (Ra)
	Ore	1.0×10^2–4.0×10^3 Bq kg^{-1} (U_{nat})
	Ore	15–1.5×10^2 Bq kg^{-1} (Th_{nat})
	Ore	6.0×10^2–3.0×10^3 Bq kg^{-1} (Ra)

Source: Adapted from IAEA (2003a).

radiological implications. Environmental modeling is a tool that can be used to decrease the radiological implications of NORM/TENOR.

5.2.2. Objective of environmental modeling

The first step in environmental modeling is to define its objectives. Examples of objectives directing an environmental modeling effort to evaluate the radiological consequences of NORM/TENORM release into the environment may be the following:

1. direct external irradiation due to NORM/TENORM-bearing air and resuspended soil;
2. inhalation of NORM/TENORM-bearing air and resuspended soil; and
3. ingestion of NORM/TENORM-bearing crops, plants, and animals.

From the practical point of view of regulations, dose limitation needs to be somewhat associated with modeling. Metabolic models are well established, at least when one considers the ICRP and the IAEA publications, and can be considered fairly general. Environmental modeling, however, can seldom be considered general enough to be applied to distinct sites. As a consequence, site-specific environmental models are quite helpful to establish measurable quantities – like concentrations in environmental gases, solids, and liquids – given in units of $Bq\,kg^{-1}$ or $Bq\,m^{-3}$.

It is well known that risk characterization for radiation exposure is usually expressed in terms of numerical probability of cancer induction in an individual, or the number of cancers in a population over a certain time interval. Fatal cancer is the end point of risk in most cases, whereas other types of cancer, not necessarily fatal, may also be considered. A relevant discussion on this subject can be found in one of the ICRP reports (ICRP, 1991).

Figure 5.6 illustrates how an assessment of risk may be associated with a given concentration of radionuclide in the environment, or vice versa.

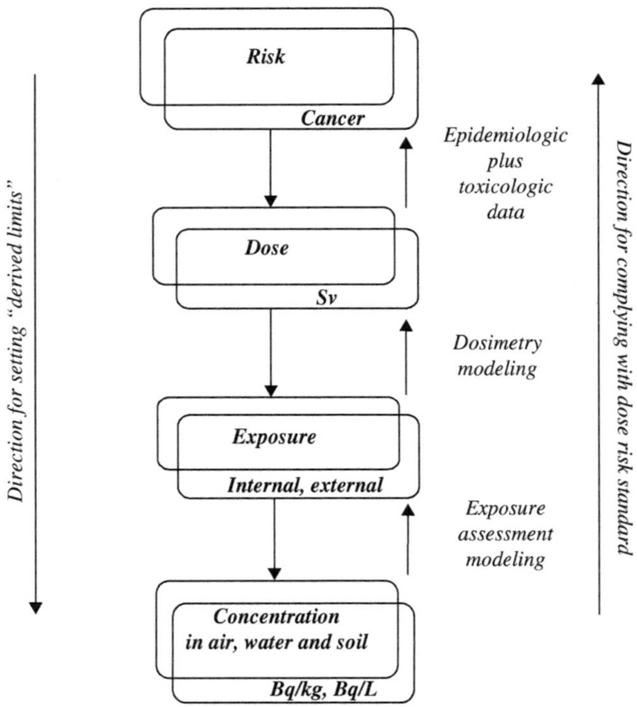

Figure 5.6 Relationships among the concepts of risk, dose, exposure, and activity concentrations in environmental media. Adapted from NAS-NRC (1999).

Typically environmental modeling is, in a very simplified way, based on a differential equation, to be solved for radionuclide concentration or dose in accordance with the following:

Source term → transport and interaction terms → concentration or dose

Figure 5.7 is a compartmental representation of a general environmental model that includes source compartment, transport and interactions compartment, and concentration and dose compartments. The source compartment can include NORM/TENORM-contaminated water, air, soil, or other material – like contaminated building material, sludge, scale, or equipment. The exposure pathways can be either direct (external or internal) or indirect through transport and interaction pathways. These exposure pathways will lead to the contaminated compartment. The model may end in this compartment or proceed further to estimate doses. Of course, water, soil, air, plants, animals, foodstuffs, and so on are included in the transport and interaction compartment.

There are a number of complications in applying such a generic structure of a model in a practical situation. Concepts like the distribution coefficient K_d and concentration factor CF are just two examples.

The structure of a model will depend on the objectives to be reached. However, it is worth mentioning that the output of a model will depend very much on the reliability of the parameters used. Ranking parameters

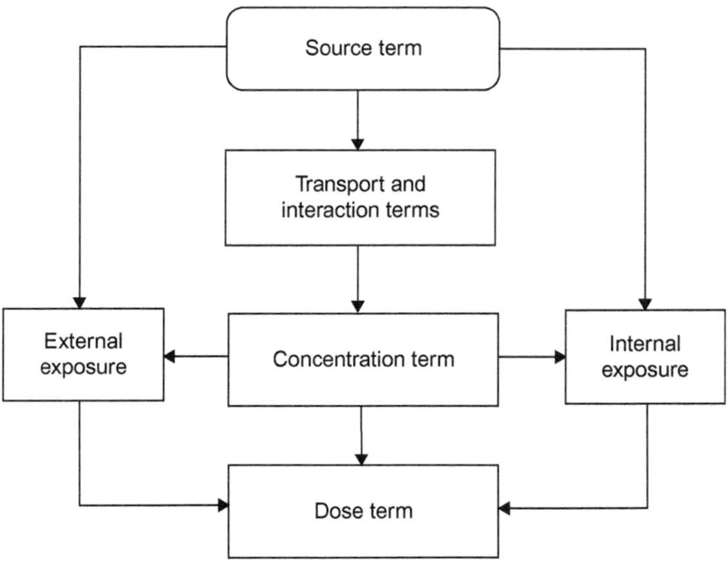

Figure 5.7 A compartmental representation of a general environmental model that includes source, transport and interactions compartment, and concentration and dose compartments.

after performing sensitivity analysis is a technique that has been used in environmental mathematical modeling (Paschoa, 1988), but it is derived from economic considerations (Tomovic and Vukobratovic, 1972).

The IAEA's EMRAS program has been quite helpful already (IAEA, 2007). Recognizing the importance of the NORM issue for most member states, the IAEA established a Working Group (WG) on modeling of naturally occurring radioactive materials (NORM) releases and of remediation benefits for sites contaminated by extraction industries. The 7th meeting of this WG occurred in Catania, Italy, May 7–10, 2007. One important remark made at the Catania meeting was that future use of NORM-contaminated/remediated sites for recreational, agricultural, industrial, or residential uses will have to be taken into account when modeling the environmental impact.

CHAPTER 6

Legal Aspects of Natural Radiation

6.1. Protection against TENORM Exposures

Workers employed by industry working with NORM and members of the public exposed to TENORM in products or waste also incur radiation doses from the natural radiation environment (NRE), either through direct irradiation from outside the body or radionuclides incorporated into the body and therefore irradiation from within.

As described previously, radionuclides of natural origin belong to two categories: (a) primordial radionuclides, naturally present in the rocks and minerals of the Earth's crust – such radionuclides of natural origin occur in significant quantities worldwide and include uranium (^{235}U, ^{238}U), thorium (^{232}Th) and their decay products, as well as potassium (^{40}K) – see also Section 6.3.1; (b) cosmogenic radionuclides created as a result of continuous interaction of cosmic radiation with target atoms in the Earth's atmosphere and in the Earth's crust, such as ^{3}H and ^{14}C (UNSCEAR, 2000).

All people are exposed to the NRE. Exposure to global average activity concentration values of the uranium and thorium series represents about 50% of the natural background radiation and contributes approximately 80% of the background dose (including radon).

The environmental activity concentration of radionuclides of natural origin varies significantly worldwide (UNSCEAR, 2000). Redistribution of materials containing NORM due to movement from the Earth's mantle to the crust, chemical interaction in the crust, sedimentation, and weathering can cause an increased concentration of natural radionuclides in some minerals and geological formations. Subsequently, mining of such ores (e.g., phosphate) or extraction of oil from underground reservoirs can lead to elevated activity concentrations in the raw material itself and wastes resulting from related industrial production processes.

Globally, hundreds of millions of tons of such NORM wastes are generated from a wide range of industries every year, such as ion exchange resins from municipal drinking water treatment, ash from coal-fired power plants, and waste water and tailings from metal mining. Although these industries deal with significant quantities of NORM, for decades many of them were not aware at all or had only insignificant knowledge about the need to implement radiation protection measures. Examples of such

Radioactivity in the Environment, Volume 17
ISSN 1569-4860, DOI 10.1016/S1569-4860(09)01706-9

protective measures cover a wide range, such as respiratory protection of workers in the oil and gas industry engaged in sand-blasting steel tanks coated with ^{226}Ra-containing slag; regulating indoor exposure of home-owners living in dwellings made of alum-shale bricks or plasterboard walls using recycled wastes with elevated ^{226}Ra concentrations from the phosphate fertilizer industry; and applying dosimetric supervision of staff working onboard high-altitude supersonic civil aircraft exposed to increased levels of cosmic radiation.

The following section describes the principles and basic requirements for radiation protection and safety applicable to all activities due to the exposure to elevated levels of natural radiation in general and exposure to TENORM during production, maintenance, and decommissioning in particular.

6.2. Heterogeneous International Approach

Although the first reports of NORM appeared in the beginning of the 20th century (McLennan, 1904 as cited in IAEA, 2003b), at the international level there was only limited and largely incoherent regulatory control over TENORM-related exposures until the 1990s. For example, significant differences existed in one and the same country between national federal agencies and others with regard to TENORM guidelines, representing different national policy judgements for risk management (NAS-NRC, 1999). This was caused primarily by different views on the maximum acceptable risk to the public. Furthermore, in countries with pertinent national regulations, this regulatory scheme was highly fragmented for example, the US Environmental Protection Agency developed TENORM standards under the *Clean Air Act* (CAA), *Clean Water Act* (CWA), *Safe Drinking Water Act* (SDWA), and *Comprehensive Environmental Response, Compensation and Liability Act* (CERCLA). Also, the local and regional variability of NRE and the resulting doses were not accounted for adequately in such regulations, exceeding by far the differences in the target regulatory level which are fractions of the natural background level.

An added level of complexity is due to the fact that the main concern in industries dealing with NORM is radium (^{226}Ra), the gaseous decay product radon (^{222}Rn) and its solid decay products, the very same radionuclides contributing most to the total radiation dose due to the NRE. This complicates the application of radiation protection concepts, since they were developed primarily for exposure to man-made radiation sources. For example, in the United States, most radionuclides are regulated under the authority of the *Atomic Energy Act* (AEA). Yet TENORM is not covered by the AEA, unless the material is specifically characterized as *source material* (i.e., uranium and thorium mill tailings, high-grade uranium and thorium ores).

The current international situation is still characterized by a significant heterogeneity among nations and their radiation protection concepts regarding NORM. The underlying problem is the lack of international agreement about the definition of NORM itself and the industrial activities dealing with NORM. This affects both regulatory approaches to products containing NORM, as well as managing NORM-containing waste. While a limited number of countries have a comprehensive system of radiation protection and waste management, many countries lack specific regulations limiting TENORM exposure of workers and consumers and addressing NORM-containing waste management today.

6.3. Regulatory Framework for NORM Industries

6.3.1. Guidance

The international community, under the leadership of the International Atomic Energy Agency (IAEA) and based on recommendations by the International Commission on Radiation Protection (ICRP), has established comprehensive guidance on TENORM-related radiation protection for NORM and TENORM, addressing the following topic areas:

1. basic requirements for radiation protection and safety principles (IAEA, 1985, 1996a, 1996b);
2. guidance on meeting the occupational protection requirements (IAEA, 1999a);
3. objectives and principles of radioactive waste management (IAEA, 1995);
4. predisposal and management of radioactive waste (IAEA, 2000a);
5. guidance for the management of NORM wastes from the mining industry (IAEA, 2000b); and
6. guidance relevant to the oil and gas industry (IAEA, 2003b, 1999b, 2000c).

All industrial operations need to fulfil some or all of the following requirements in terms of radiation protection concerning TENORM: (i) control over occupational radiation exposures of workers dealing with NORM; (ii) radiation protection of members of the public through the proper management of radioactive wastes containing NORM; and (iii) safe management of the decommissioning of industrial facilities previously using NORM. Radiation protection measures for the aviation industry exposed to elevated levels of cosmic radiation are addressed in a separate paragraph below.

At the international level, a national competent agency with the authority to establish or adopt regulations is charged with the responsibility to establish a process for authorizing an industrial activity involving NORM. Typically, this agency establishes radiation protection standards (e.g., setting occupational dose limits or dose limits for members of the public). Also, such an agency has the authority to establish mechanisms for exclusion, exemption, or clearance from regulatory control (see Section 6.3.2.3 on *exemption*, *clearance*, and *authorized release*).

In order to decide how to regulate industrial activities involving NORM, the current radiation protection concept differentiates between two largely different activities, that is, *practice* and *intervention*:

7. *Practice* encompasses the following human activities: (a) introducing additional sources of radiation exposure or exposure pathways; (b) extending radiation exposure to additional people; (c) modifying the network of exposure pathways from existing sources, so as to increase the exposure or the probability of radiation exposure of people or the number of people exposed.
8. *Intervention* is any action taken with the intention to (a) reduce radiation exposure, (b) avert radiation exposure, and (c) reduce or avert the probability of radiation exposure. In either case, these actions are focusing on exposure to radiation sources which are not part of a controllable practice (OBS: *Intervention* is also applicable to radiation sources which are out of control due to an accident).

The common radiation protection goal in all NORM-related situations is to meet two requirements: (i) keep radiation doses as low as reasonably achievable, economic and social factors being taken into account (*ALARA Principle*); (ii) ensure that radiation doses are below the regulatory dose limits for workers and members of the public.

Workers require radiation protection during their occupational exposure to NORM occurring predominantly during mining of minerals and extraction of oil and gas, concentration, processing of the raw material, and onsite handling/storage of waste.

Members of the public can be exposed to TENORM through the airborne release of NORM particulates and radon or groundwater contamination from wastes disposed onsite in piles. Also, inadequate waste disposal, recycling of waste material (e.g., backfill underneath dwellings; concrete aggregate), and consumer products can cause elevated radiation exposure. Although levels of TENORM exposure vary widely, not all of these exposure situations are subject to regulatory control. For example, some mining activities extract ore with highly increased uranium and/or thorium activity concentrations; such activities are considered *practices*. However, if the radiation exposure situation is not amenable to control through the requirements of the radiation protection system (e.g., solar flares irradiating astronauts during manned space flights or exposure due to unmodified concentrations of radionuclides in most raw materials), such exposure is deemed to be excluded from radiation protection standards.

6.3.2. Practical implementation

The following section describes the widely accepted regulatory approach developed by the European Union (EU)-EURATOM and the IAEA, jointly sponsored by the Food and Agriculture Organization of the United Nations (FAO), the International Labour Organization (ILO), the Nuclear Energy Agency (NEA) of the Organisation for Economic Co-operation and Development (OECD), the Pan American Health Organization (PAHO), and the WHO – see IAEA (1996a).

6.3.2.1. NORM workers

The EU Directive Title VII of the BSS Directive (EC, 1997) applies to work activities within which the presence of natural radiation sources leads to a significant increase in the exposure of workers or of members of the public which cannot be disregarded from the radiation protection point of view. It requires Member States to identify work activities that may be of concern, such as work activities involving operations with NORM and aircraft operation (EURATOM, 1996; EC, 1996b). It also addresses work-related exposure to radon and thoron decay products or external gamma radiation.

NORM-relevant activities, subject to this EU Directive, are further specified as including (EC, 1997):

1. work activities involving operations with, and storage of, materials not usually regarded as radioactive but which contain naturally occurring radionuclides, causing a significant increase in the exposure of workers and, where appropriate, members of the public;
2. work activities which lead to the production of residues not usually regarded as radioactive but which contain naturally occurring radionuclides, causing a significant increase in the exposure of members of the public and, where appropriate, workers.

Already before the BSS Directive came into force in the Member States on May 13, 2000, NORM issues were studied in several EU Member States (EC, 1996a). It became evident that a major unresolved issue is when to require regulatory control for radioactive effluents resulting from industries dealing with NORM. In particular, there are diverging views on setting a NORM-related threshold dose value in all Member States, with discussions on such a criterion ranging from 10 to 300 $\mu Sv\,y^{-1}$ (EC, 2004).

A group of experts, established under the EURATOM Treaty, proposed a four-band system with marker points between each for regulatory control above the limit of 1 $mSv\,y^{-1}$. As far as regulatory control is concerned, the band structure is as follows:

- band 1 – no need for regulatory control (below an exemption level);
- band 2 – lower level of regulation;
- band 3 – higher level of regulation; and
- band 4 – not to be permitted without full individual assessment.

Taking into account the identified pathways for each case, screening and reference levels were derived to indicate the likely level of regulation for the radionuclides of concern to the relevant industries. Thus, for example, if the derived screening and reference level put the industrial process in band 1, there would be no need for regulatory control. However, if the derived level placed the industrial process above band 1, control measures would be applied or the process should receive individual assessment prior to deciding which control measures to adopt (EC, 1997).

Taking into account appropriate exposure pathways in conjunction with the EURATOM system of bands, screening concentration levels can be established for ^{226}Ra plus progeny, ranging from 30 to 300 Bq g^{-1}. Thus, for concentrations below 30 Bq g^{-1}, there would be no need to consider regulatory control and, above 300 Bq g^{-1}, individual assessment would be necessary prior to deciding which control measures to take.

Those screening concentration levels are also in accordance with the concept favored by the US National Academy of Sciences – National Research Council on Naturally Occurring Radioactive Material, which suggests a relationship between the concentration of radionuclides in the environment, exposure, dose, and resulting risk (NAS-NRC, 1999). The underlying concept assumes cancer mortality, or in some cases cancer incidence (morbidity), as a potential measure of risk due to exposure to radiation. However, in many environmental radiation assessments, the endpoint of the assessment is dose rather than cancer risk. In such a case, an assessment of the risk corresponding to a given concentration of various environmental radionuclides can be used instead, or vice versa.

In a further development, secondary levels of dose and derived limits could be adopted, for example, for ^{226}Ra releases, as proposed in Figure 6.1. The sequence of dose-related levels and calculated derived limits shown in Figure 6.1 can be used in decision-making processes involving TENORM. To do so, once a primary dose limit is established, a fraction of this primary dose unit can be chosen after taking into account societal and economic considerations. Thus, if a registration level is established near the highest value of the variation of the natural radioactive background, investigational, intervention, and remedial action levels can be defined above such a level. The levels shown in Figure 6.1 are defined in terms of fractions of a primary dose but can be implemented practically in terms of units, for example, in Bq m^{-3} for water, air, and other gases; Bq kg^{-1} for pasture, soil, or scrap metals; and Bq y^{-1} for rates of discharge. Derived limits can be calculated using environmental models, thereby allowing for a sensitivity analysis for ranking the importance of individual model parameters.

In view of the lack of an international agreement on a practically applicable approach on regulatory TENORM control, individual activities have been initiated to arrive at a coordinated concept. Such an example is the interactive

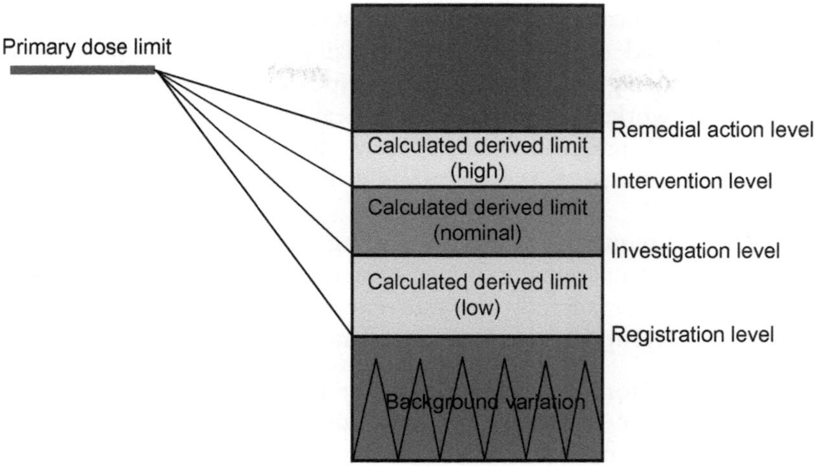

Figure 6.1 Schematic representation of calculated derived limits (low, nominal, and high) in accordance with parameters used in an environmental model calculation plus registration, investigation, intervention, and remedial action levels in comparison with the natural radiation background variation. (Paschoa, 2003)

network (NORMnet), operated on behalf of the European Commission, Radiation Protection Unit, since 2005. The objective of this website is to help NORM stakeholders to communicate, share knowledge, identify problems, and propose solutions to numerous issues in the industry. See: http://europa.eu.int/comm/energy/nuclear/radioprotection/index_en.htm; registration by e-mail to Kathy.Hillis@nnc.co.uk.

6.3.2.2. Civil aviation crew members

Already in 1991, the ICRP had recommended that aircrews working in civil aviation should be considered as occupationally exposed radiation workers due to their exposure to cosmic radiation (ICRP, 1991). It provided further guidance in 1997, supporting this approach (ICRP, 1997b).

The civil aviation industry is addressed specifically in Title VII of this EU Directive. In 1996, the European Commission required all Member States to introduce national legislation to assess and regulate radiation exposure received by crews working aboard civil aircraft by May 2000 (Article 42 of EURATOM, 1996). It is interesting to note that a need for justification is not explicitly stated. This Directive requests each Member State to take the following measures:

3. assess the radiation exposure of the crew concerned;
4. take account of the exposure of aircrew whose dose might exceed 1 mSv y^{-1};

5. although optimization is not explicitly stated, employers are to account for the assessed exposure when organizing working schedules with a view to reducing the doses of highly exposed aircrew;
6. inform the workers concerned of the health risks their work involves;
7. in the case of a pregnant female aircrew, apply the ALARA Principle for the equivalent dose to the foetus/child. However, there are no explicit limits;
8. there are no requirements for designating areas or categorizing workers, contrary to occupational radiation protection approaches; and
9. dose records for aircrew likely to exceed $6\,\text{mSv}\,\text{y}^{-1}$ are to be kept, foreseeing appropriate medical surveillance.

Since this sector of industry has never been subject to any radiation protection scheme hitherto, the introduction of this scheme caused significant operational changes and associated costs, such as reducing the flying hours of individual crew members, using lower flight altitudes, and rotating crew members on routes associated with higher radiation doses with members receiving lower doses.

6.3.2.3. Exemption, clearance, and authorized release

Radiation protection concepts have been developed for practices and sources which are likely to result in only minor radiological impact on man and the environment, that is, *the concept of exemption*, *the concept of clearance*, and *the concept of authorized release*. It is emphasized that these concepts are not intended for practices which would otherwise not be justified.

Regulatory authorities may exempt practices and sources within practices from notification, registration, or licensing, if (a) the sources meet the exemption criteria, (b) the sources meet the exemption levels specified by the authority on the basis of these exemption criteria. The principles for exemption are defined as (IAEA, 1988):

1. The radiation risks to individuals caused by the exempted practice or source are too low to be of any regulatory concern.
2. The collective radiological impact of the exempted practice or source is sufficiently low not to warrant regulatory control under the prevailing circumstances.
3. The exempted practices or sources are inherently safe, without any appreciable probability of scenarios that could lead to a failure to meet the two principles listed above.

A practice or a source within a practice may be exempted from regulatory control if the following criteria are met in all feasible situations: (a) the effective dose expected to be incurred by any member of the public due to the exempted practice or source is of the order of $10\,\mu\text{Sv}$ or less in a year and (b) either the collective effective dose committed by one year of performance of the practice is no more than about 1 man-Sv or an assessment for the optimization of radiation protection shows that exemption is the optimum option. Here, it must be observed that the application of the concept of exemption to natural radionuclides, where these are not excluded, is limited to the incorporation of naturally occurring radionuclides into consumer products or their use as a radioactive source (e.g., ^{226}Ra, ^{210}Po) or for their elemental properties (e.g., thorium, uranium).

In the European Union, some Member States have discussed the benefits of a concept consisting of reporting and prior authorization and its corollaries, exemption and clearance, laid down in Title III of the Directive for practices, to work activities with regard to natural exposures as well (Henrich and Janssens, 2005). Practical guidance on the application of these concepts and clearance levels has been developed for Member States (EC, 2002).

Some NORM-related radiation exposures can meet the requirements of the *concept of exemption* from regulatory control, if the component of exposure is not amenable to control by reasonable practical means. Such an approach could be useful for recycling relatively small amounts of NORM wastes with very low levels of contamination, for example, as construction material for roads. The difficulty lies in the definition of *reasonable*, since this will depend on the technical and financial means of the society willing to engage in such recycling activities. A sound baseline for deciding on exemption could be a level of radiation-induced health risk considered too trivial to control, since the cost associated with the control would be higher than *any* benefit resulting from the regulatory control. Applying the criteria foreseen in the IAEA BSS for the exemption of a practice or a source within a practice (IAEA, 1996a) to such a recycling program, this would translate to an effective dose to a member of the public of 10 μSv or less in a year to be exempted. Alternatively, the collective effective dose of 1 man-Sv committed by one year of using or being exposed to the waste would also be exempted from regulatory control. Following the EU BSS, a similar approach for deriving exemption values could be envisaged (EURATOM, 1996). The scenarios considered to determine annual individual exposure from exempted sources could account for normal use, misuse, and disposal of the wastes.

For the treatment of the predominantly large amounts of very low level NORM waste, the *concept of clearance* has a certain attraction. This concept foresees clearance of radioactive materials from regulatory control at or below a certain level of activity concentration. It requires the predetermination of *clearance levels* for radionuclides from a prescribed dose limit. Such a process should assume realistic parameters for establishing the different exposure pathways, for example, definition of complex exposure pathways

Table 6.1 Recommended values for limiting radiation exposure.

Exposure group	Recommended values	Comments
Occupational exposure	20 mSv y^{-1}	Effective dose averaged over five consecutive years; the start of the averaging period shall be coincident with the first day of the relevant annual period after the date of entry into force of the standards, with no retroactive averaging
Occupational exposure	50 mSv	Effective dose in any single year
Occupational exposure to radon	1.0×10^3 Bq m^{-3} 6×10^6 Bq h m^{-3}	The International Commission on Radiological Protection has recommended that action levels for occupational exposure to radon can fall in the range 500–1,500 Bq m^{-3} (ICRP, 1993)
Occupational exposure for apprentices (16–18 years of age)	6 mSv 2×10^6 Bq h m^{-3}	Effective dose in a year for persons who are training for employment involving exposure to radiation Radon exposure
Occupational exposure under special circumstances	20–50 mSv y^{-1}	Dose averaging period may exceptionally be up to 10 consecutive years
Occupational exposure to natural radiation sources	20 mSv y^{-1}	Work activities within which the presence of natural radiation sources leads to a significant increase in the exposure of workers
Public exposure to radiation sources	1 mSv y^{-1}	Estimated average doses to the relevant critical groups of members of the public that are attributable to practices, in addition to natural radiation exposure
Public exposure under special circumstances	5 mSv	Effective dose in a single year provided that the average dose over five consecutive years does not exceed 1 mSv y^{-1}
Public exposure to radon	Up to 1,000 Bq m^{-3}	Values vary widely between recommended limits and legally binding values in different countries

Table 6.1. *(Continued)*

Exposure group	Recommended values	Comments
Public exposure to natural radiation sources	$10\ \mu Sv\ y^{-1}$	Effective dose or derived concentration values based on this dose value can be considered as trivial and therefore need not be subject of regulatory control
	Below $0.2\ Bq\ g^{-1}$	Activity concentration for a nuclide of the uranium or thorium decay chain can be considered as trivial and therefore need not be subject of regulatory control

versus enveloping scenarios (including inhalation, ingestion, and external radiation). It is important to bear in mind that a clearance level is a *reference level* and not a *limit.* This concept is different from the exemption concept: the radioactive materials subject to clearance are already under regulatory control until cleared by the authority. Such material is outside the regulatory system. This concept has already been applied in decommissioning and dismantling of nuclear installations with significant success in Spain, Germany, and, to a more limited extent, in the United Kingdom and Belgium.

An *authorized release* of a building or a site contaminated with NORM ensures that all future uses will result in doses no higher than an annual dose constraint. This has the inherent advantage of enabling the prediction, with a low degree of uncertainty, of the maximum radiation dose to persons who will use the cleaned up and restored area in the future. It also includes the option for regulatory authorities to issue a *release for restricted use.* By prescribing how to use the area still affected (to a limited extent) by residual radioactivity after NORM deployment will ensure the public that an annual dose constraint will not be exceeded, as long as the user adheres to the restrictions.

Table 6.1 contains a survey of recommended values for exposures attributable to *practices*, with the exception of exposures from natural sources that cannot reasonably be regarded as being under the responsibility of any principal party involved in radiation protection.

CHAPTER 7

Terrorism and Natural Radiation

7.1. Natural Radionuclides as a Terrorist Weapon

The subject of nuclear terror has been discussed in a number of published and unpublished studies, before and after the tragic 9/11 episode in the United States. Radiological terror, on the other hand, has not received as much attention, with the exception of the so-called radiological dispersion devices (RDDs).

Terrorism aims to induce fear and uncertainty, inflict casualties, disrupt society, and cause economic loss and question the legitimacy of government in order to induce political change. Risk assessment for an act of terror includes the assessment of the probability of occurrence of such an act, multiplied by the damage caused if the incident occurs. This risk can be modified by the countermeasures society is able to take in order to reduce the consequences induced by the terror attack. All of this also applies to terror attacks using natural radioactivity, that is, intentional contamination with radioactive material might be combined with other terrorist acts, such as the deployment of an *improvised explosive device* (IED). Even if the public should be getting more or less used to traditional terrorism, using explosives, guns, and knives in order to threaten, maim, or kill, the use of radioactivity will certainly add another layer of fear. The international scientific community, together with security forces and intelligence services, is increasingly concerned about terrorist groups interested in acquiring radioactive material to be used as a radiological weapon. Although few deaths are to be anticipated from the radiation exposure, members of the public will be affected by such a terror attack much more than by a conventional attack: the destruction resulting from an IED, combined with the widespread fear of anything radioactive, is likely to increase significantly the psychological impact induced by the terror attack (The Bratislava Report, 2009).

Therefore, authorities should plan and practise the issuing of immediate substantive instructions to the public, as well as comprehensive mitigation and cleanup procedures. In order to assist the authorities with the preparatory work and the assessment of the likelihood for such an attack, this section reviews the objectives, logistical and technical prerequisites for terrorists to engage in such activities using natural radionuclides.

Radioactivity in the Environment, Volume 17
ISSN 1569-4860, DOI 10.1016/S1569-4860(09)01707-0

The objectives of this analysis are:

(a) improved understanding of possible terrorists attack modes;
(b) provision of assistance to the members of the security community in assessing the significance of suspicious activities, thereby pre-empting any planned nuclear terrorism activities.

OBS: For reasons of security, technical details will be omitted.

7.2. Suitable Natural Radionuclides

Natural radioactivity is, as noted before, of two types, the primordials like uranium (^{238}U) and the cosmogenics like tritium (^{3}H). Although tritium and polonium (^{210}Po) are constantly produced naturally in the atmosphere, commercially available ^{3}H and ^{210}Po are largely man-made in reactors or accelerators. This chapter focuses on the primordial type of natural radioactivity.

When one mentions radiological terrorism, an RDD comes usually to the mind. However, as has already been explained, a radiological attack can be either active or passive (Steinhäusler et al., 2008). The choice of radionuclide to be used depends on the type of attack that the terrorist intends to deflagrate. Table 7.1 lists a selection of radionuclides considered adequate for radiological attacks. It must be mentioned that there is quite a variety of radionuclides which can also be used for terror attacks, but most of those listed in Table 7.1 are fairly easy to obtain either in hospitals or in research and industrial laboratories.

Figure 7.1 is a graph of the dose coefficient ($Sv\,Bq^{-1}$) as a function of radionuclide. One can see from Table 7.1 and Figure 7.1 that the radionuclides ^{226}Ra, ^{210}Po, and ^{241}Am have dose coefficients which are three times higher than all the others. This means that these three radionuclides have higher dose effectiveness per Becquerel ingested. Other radionuclides like ^{252}Cf, ^{90}Sr, and ^{137}Cs also deliver high dose effectiveness per Becquerel ingested, but to a lesser extent.

For a terror attempt which would involve ingesting a radionuclide, the three radionuclides of choice would be ^{241}Am, ^{210}Po, or ^{226}Ra. The final choice would depend on how easy or difficult it would be to obtain one of these three selected alpha emitters. Thus, if a terrorist (or a group of terrorists) has access to an amount of a radionuclide which can deliver doses many times higher than the dose limit for a particular radionuclide, the way to deliver a fatal dose by means of ingesting a particular radionuclide becomes a matter of choice and capacity to obtain the radionuclide.

Those radionuclides which can be used in radiological attacks per body burdens necessary to deliver fatal doses as well as the mass of each radionuclide associated with the respective fatal body burden would be ranked as

Table 7.1 A selection of radionuclides with potential to be used in terror attacks (data from Paschoa and Dantas, 2008).

Radionuclide		Half-life (years)	Specific activity (GBq g^{-1})	Dose coefficients[a] (Sv Bq^{-1})		Source
				Inhalation[b]	Ingestion	
1	^{241}Am	432	1.18×10^{2}	3.9×10^{-5}	2.0×10^{-7}	Lab, Ind
2	^{252}Cf	898	2.4×10^{4}	8.3×10^{-6}	4.3×10^{-8}	Lab
3	^{137}Cs	30	3.6×10^{3}	4.8×10^{-9}	1.3×10^{-8}	Lab, Hosp, Ind
4	^{60}Co	5.3	4.1×10^{4}	9.6×10^{-9}	3.4×10^{-9}	Lab, Hosp, Ind
5	^{125}I	0.16	6.3×10^{5}	1.4×10^{-8}	1.5×10^{-8}	Hosp
6	^{131}I	2.2×10^{-2}	4.4×10^{6}	2.0×10^{-8}	2.2×10^{-8}	Lab, Hosp
7	^{192}Ir	0.20	3.4×10^{5}	6.3×10^{-9}	1.4×10^{-9}	Lab, Hosp
8	^{32}P	3.9×10^{-2}	1.1×10^{7}	3.2×10^{-9}	2.3×10^{-10}	Lab
9	^{210}Po	0.38	1.7×10^{5}	3.9×10^{-6}	2.4×10^{-7}	Lab, Ind
10	^{226}Ra	1.6×10^{3}	37	3.2×10^{-6}	2.8×10^{-7}	Lab, Ind
11	^{90}Sr	29	5.6×10^{3}	1.5×10^{-7}	2.8×10^{-8}	Lab
12	^{99m}Tc	6.8×10^{-4}	1.9×10^{8}	1.2×10^{-11}	2.2×10^{-11}	Hosp
13	^{99}Mo	7.5×10^{-3}	1.7×10^{7}	9.7×10^{-10}	1.3×10^{-9}	Lab

Sources: Based on ICRP Publication 38 (ICRP, 1983); http://www.iem-inc.com.toospar.html
Notes: Lab, laboratory; Ind, industry; Hosp, hospital
[a]IAEA Safety Guide RS-G-1.2 (IAEA, 1999c).
[b]All except ^{125}I and ^{131}I have AMAD = 1 μm.

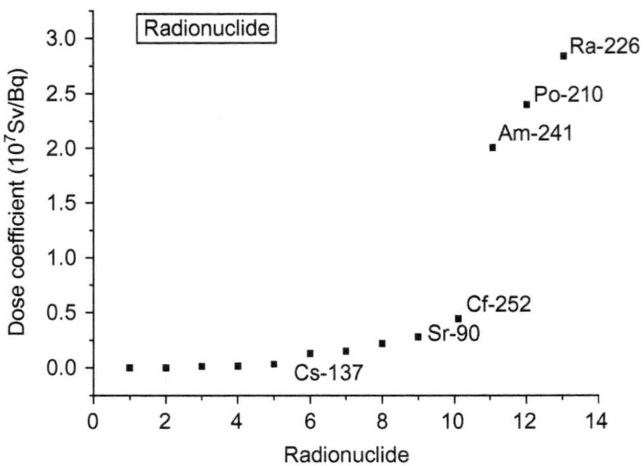

Figure 7.1 Graph of dose coefficient ($\times 10^7$ Sv Bq^{-1}) as a function of radionuclide.

Table 7.2 Characteristic properties of natural radionuclides suitable for a radiological terror attack.

^{226}Ra	^{210}Po
Long half-life: 1,600 years Low specific activity: 37 GBq g^{-1} Energy of alpha emission: 4.8 MeV Available as hospital and laboratory source	Short half-life; 140 days High specific activity: 170 TBq g^{-1} Energy of alpha emission: 5.3 MeV Available as laboratory source

$^{241}Am < ^{210}Po < ^{226}Ra < ^{252}Cf < ^{90}Sr < ^{137}Cs$ – or by the mass needed to attain the fatal dose: $^{210}Po < ^{252}Cf < ^{90}Sr < ^{137}Cs < ^{241}Am < ^{226}Ra$. Considering together these two rankings, ^{210}Po presents the intrinsic advantage of small body burden and small mass to achieve a fatal result.

Of all the many radionuclides commercially available, only polonium (^{210}Po) and radium (^{226}Ra) are suitable for a terror attack. The very different radiological characteristics of these nuclides are summarized in Table 7.2.

Radium, chemically similar to barium, is found in pitchblende at about 0.14 g Ra ton^{-1}. Polonium, chemically similar to tellurium, a silvery, shining metal, is one of the rarest elements.

In radiological terms, both radionuclides are alpha emitters of similar energy, but have very different specific activity: polonium exceeds that of radium by almost a factor of 5,000.

7.3. Illegal Acquisition of Natural Radioactive Material

The natural radionuclides radium (^{226}Ra) and polonium (^{210}Po) have been subject to trafficking worldwide, together with natural uranium ore (^{238}U), processed uranium (*yellow cake*), low-enriched uranium (<20% ^{235}U), highly enriched uranium (>20% ^{235}U), and natural thorium ore (^{232}Th). Figure 7.2 shows that ^{226}Ra (mostly hospital sources) represents a non-negligible portion of the radioactive material trafficked and detected; trafficking ^{210}Po is a singularity so far.

The attack on the former Russian KGB agent Alexander Litvinenko with ^{210}Po poisoning led to his death on November 24, 2006, (see, e.g., among other sites, http://www.pakistanlink.com/Commentary/2007/Jan07/05/02.HTM). No other such case with proven use of a natural radionuclide with criminal intent is known to the authors (i.e., Litvinenko was the first person to die from acute exposure to ^{210}Po alpha radiation).

As a matter of fact, the first victim of polonium radiation effects was Nobel Prize winner Irene Joliot Curie who died in 1956 from leukemia

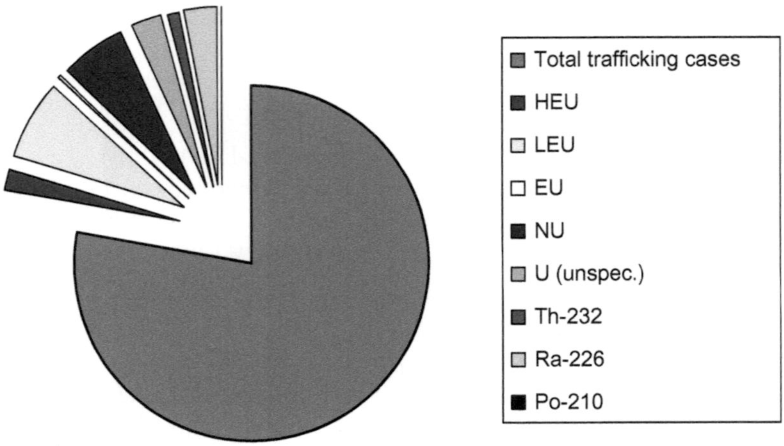

Figure 7.2 Global trafficking of radionuclides (DSTO 1991–2008).

attributed to her alpha radiation exposure when a polonium source exploded in her laboratory a decade earlier (see, e.g., Q&A: Polonium 210, 2006).

Illicit trafficking of natural radionuclides and all associated operational procedures can be subdivided into five phases:

- single or repeated diversion of the material from the site of production or storage, or during transport;
- covert transport of the diverted material to a third location in order to trade it on the black market;
- searching for potential buyers of the diverted material, using only small samples of the material offered for sale; alternatively, phase 3 is not needed if the diversion was ordered beforehand by the future buyer;
- closing of the sale with the buyer, typically associated with quality control of samples; and
- covert delivery of the total amount of material.

Any action aimed at interfering with illicit trafficking will need to address all of these stages, that is, strengthen the national infrastructure for material protection and accounting, develop higher standards of good governance, and improve the control system deployed by customs, border guards, and security forces.

7.4. Motivation for a Terrorist Attack with Natural Radionuclides

Radiological attacks can occur for different motives, ranging from personal grudges against colleagues at work and family members (i.e., criminal incidents) to political blackmail against the authorities (i.e., acts of terrorism). While several criminal incidents involving radioactive materials have occurred in the past, only a few such incidents with a terrorist background have been recorded so far. For example, the *Database on Nuclear Smuggling, Theft and Orphan Radiation Sources (DSTO)* records a total of 33 malevolent radiological acts, of which only 3 are related to terrorism (Steinhäusler and Zaitseva, 2007). So far, the motivation to use a natural radionuclide for a criminal act is limited to a probably politically motivated poisoning of a man with ^{210}Po in 2006 – as mentioned above, Alexander Valterovich Litvinenko, a former officer of the Russian State security service (FSB), and later a Russian dissident and writer, died in London due to ^{210}Po poisoning on November 24, 2006: http://en.wikipedia.org/wiki/Alexander_Litvinenko (last visited on January 8, 2009).

In summary, hitherto terrorists have not been motivated to deploy a natural radionuclide in an attack.

7.5. Modes of Attack with Natural Radionuclides

Experience of radiological attacks using natural radionuclides has been gained largely from such attacks using man-made radionuclides. Since the logistical and operational requirements are largely identical, this information can be extrapolated to terror attacks with natural radionuclides. Such terror attacks can occur in an active and passive form:

(a) Active radiological attacks through an RDD, such as
- Generation of a radioactive solution, resulting in extensive environmental contamination but low individual doses (mostly through external contamination).
- Generation of radioactive aerosols, resulting in a large collective dose (mostly through inhalation).
- Combination of radionuclide with conventional explosives (dirty bomb), likely to result in a large collective dose depending on source strength and dispersal:

> In December 1998 Chechen authorities foiled a possible radiological terrorist act when they found and defused a mine attached to a container "full of radioactive substances" near the town of Argun, located about 15 km east of Grozny (in DSTO).

(b) Passive radiological attacks through a radiation emitting device (RED) or intentional contamination of foodstuff, such as

- Covert irradiation of one or more individuals by placing a strong source in a high traffic density area

In May 2002, a nuclear expert working in a hospital in China's Guangdong Province tried to kill his colleague by installing an iridium-192 source in the ceiling of his office. The radioactive material caused illness in the man and 73 other staff members in the hospital, before it was eventually discovered and removed (in DSTO).

- Deliberate contamination of food or drinks:

In September 1998, a graduate student in Providence, Rhode Island, tried to poison his ex-girlfriend by tainting her food with iodine-125 he had stolen from Brown University's molecular pharmacology laboratory. Another male graduate student in Taiwan was poisoned by phosphorus-32 placed in his food and drink on about 30 occasions between 1 October 1994 and 15 February 1996, although not all of them involved the radioactive isotope. The substance, also stolen from a university molecular biology lab, was placed in the victim's drinking cup and on eating utensils in his workplace, the Institute of Plant Pathology, by a fellow student (in DSTO). In 1995, 26 employees of the National Institutes of Health in Bethesda, Maryland, were exposed to radioactive phosphorus, which someone had used to contaminate a water cooler. The FBI and other investigating agencies were never able to identify the culprit (in DSTO).

7.5.1. Contamination of drinking water

Natural radionuclides can be used for radioactive contamination of drinking water. Such an attack can occur on a public water supply system, both in its entirety and/or as a part of the distribution system. In the case of large water systems, for example, a major city with a drinking water reservoir, it is difficult to conceive of introducing sufficient radioactive material to render the water system a potential source of acute radiation poisoning. However, it is much more plausible to contaminate the water supply to a level considered unsafe for chronic consumption.

Radiological contamination introduced *prior* to water entering the distribution system (e.g., contamination of wells or supply canals) is likely be detected by routine public water supply monitoring and mitigated in the water treatment process.

A major concern is the introduction of contamination to the water supply *post* treatment and monitoring (i.e., radioactive material introduced into the water after it has entered the distribution system). This type of contamination would be difficult to detect and hard to remove but quite easy to accomplish with only a low probability of timely detection. Water distribution systems are maintained at a positive pressure. This allows the provision of drinking water to the upper floors of multistory buildings and adequate water pressure for firefighting. This positive pressure prevents natural contaminants from entering the distribution system at the point of a leak because the net flow is outwards. However, any building connected to the public water supply is also a potential location for terrorists to introduce radiological contamination by applying a local overpressure in the distribution system. The following components would be required for such an attack (all the mechanical items mentioned below, except the second one, are available at any hardware supply store):

- compressor or other source of pressure;
- water-soluble radioactive substance;
- few pipes and valves; and
- large hydrostatic pressure tank.

External pressure is increased to the hydrostatic pressure tank containing the radioactive solution, until the pressure in the tank exceeds the pressure in the water distribution system. Once this pressure gradient is obtained, valves are opened and the radioactive contaminants flow from the tank through the water service line connecting the building to the general water distribution system. From there the radioactive solution spreads as water is used by consumers in the neighborhood.

7.5.2. Contamination of roads

Another mode of attack using a radioactive solution aims at large-scale environmental radioactive contamination, preferably in an urban area of high commercial value. This requires radionuclides to be dissolved in large containers (e.g., 200–500 L of water in oil drums to which a drain valve has been added). The drums are located inside a truck or van. At the end of a forecast rainy period, terrorists can cause widespread environmental contamination by spreading the solution (e.g., in a commercial district). If the spraying of a radioactive solution occurs at the onset of rain, some of the radioactive solution will be washed into the sewer system.

The spraying of the solution, a small stream of radioactively contaminated water coming from the terrorists' vehicle, will not be noticed: the undercarriages of *all* vehicles participating in the traffic on a rainy day will be dripping water. Further dispersion of the radioactive solution deposited on the road will be provided by the traffic (tyre spray). In this manner, large urban areas can be contaminated relatively quickly. Once the planned roads have been contaminated, terrorists can abandon the vehicle, leave the vicinity of the contaminated area, and inform the media.

7.5.3. Generation of radioactive aerosols by aircraft and handheld devices

For the generation of radioactive aerosols, radioactive materials are selected that preferably have a high specific activity. Suitable natural radioactive material, such as $RaCl_2$ solution, is dispersed in the environment. This could be achieved with an adapted crop-dusting plane or commercially available mosquito-control fogging apparatus or even an ultralight plane especially adapted for the mission. As a result of this spraying, contamination of the environment and – dependent on the source strength and exposure conditions – considerable collective inhalation dose will result. Since the terror attacks on September 11, 2001 in the United States, security has been increased in an attempt to deny terrorists access to virtually all aircraft having a potential to do harm. However, general aviation and ultralight planes can still be leased with relative ease and could be adapted for criminal purposes at a remotely located minor or even private airfield with lower operational security.

7.5.4. Indoor contamination with ^{210}Po aerosols

Generally, aerosol delivery systems have been well covered in the literature with aerosol now including liquid droplets, solid particles, and combinations of both. There is, however, one radioactive aerosol delivery system which is dependent upon the physical property of ^{210}Po. Polonium has a melting point of 254 °C and a boiling point of 962 °C. Nevertheless, ^{210}Po has the ability to evaporate or become airborne rather easily: 50% of a ^{210}Po sample heated to a temperature of 55 °C will vaporize within 45 h. It has been hypothesized that small clusters of polonium atoms are spalled off by the alpha decay, a process similar to the recoiling of the nucleus following alpha-particle emission (see, e.g., Polonium, http://en.wikipedia.org/wiki/Polonium). Therefore, a 40-W incandescent light bulb would produce sufficient heat to disperse ^{210}Po aerosols into a room when the occupants have the lights on.

7.5.5. Dirty bomb

A *dirty bomb* is an RDD which disperses radioactivity by means of a conventional explosive propellant. This form of unconventional weapon, using natural radioactive material, might be used by terrorists to create disruption in a society through fear, actual physical harm, and economic damage. The dirty bomb mode of attack has not been utilized to date, but it has been attempted in Chechnya already, as noted above. In the following section, different attack scenarios are modeled (Steinhäusler et al., 2008).

7.5.5.1. Uranium-based RDD

One of the most important selection criteria for the radiological part of a dirty bomb is the activity–mass relationship (*specific activity*, expressed in $Bq\,g^{-1}$). Uranium, either naturally occurring or enriched to weapons grade (90% or greater ^{235}U), has a low specific activity. For ^{235}U, the specific activity value is $7.99 \times 10^{-5}\,Bq\,g^{-1}$. A mass of 25 kg of 90% ^{235}U is a sufficient amount of material for a gun-type assembly nuclear weapon and 50 kg for a bare-sphere critical mass of HEU (NAS-NRC, 1989). A *subcritical mass* (two pieces, total mass 25 kg) would represent a radiological source of about 2 GBq. Such a level of activity, even when in one piece, is hard to detect by monitoring activity. If dispersed by an explosive device, it would become much more difficult to detect. Figure 7.3 shows the dose iso-curves resulting from the detonation of a uranium-based dirty bomb (activity ^{235}U: 2 GBq), detonated by about 50 kg of explosives. Even in the immediate vicinity of the attack ($0.001\,km^2$), maximum dose values received after 1 h would not exceed 10 μSv. Due to the low specific activity of ^{238}U, the result of blowing up a truck load of yellowcake would appear to be largely psychological. The radiotoxicity of natural uranium is low

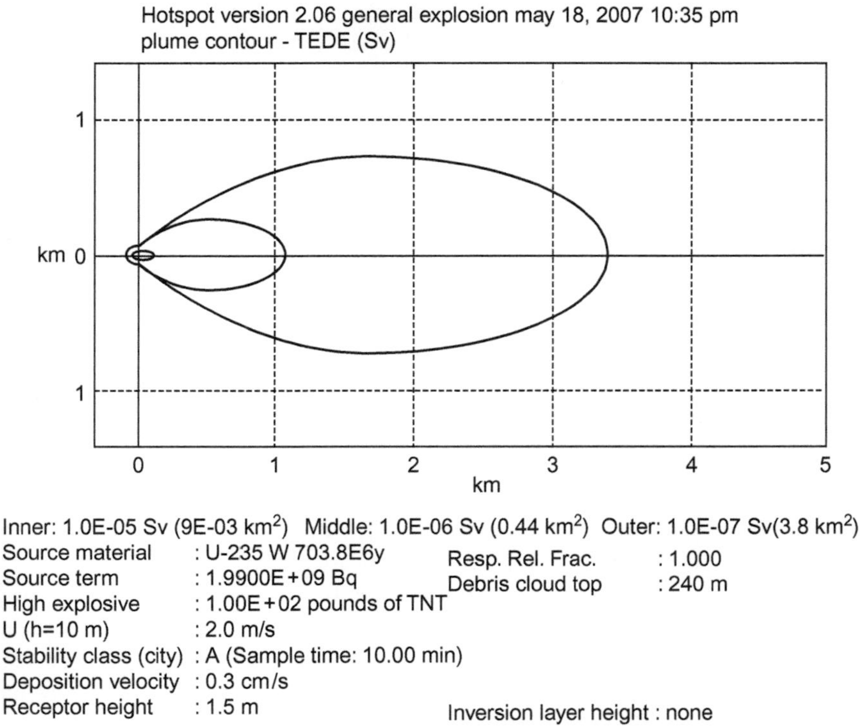

Figure 7.3 Uranium-based *dirty bomb* TEDE-contour plot in an urban environment (^{235}U-235 activity: 2 GBq).

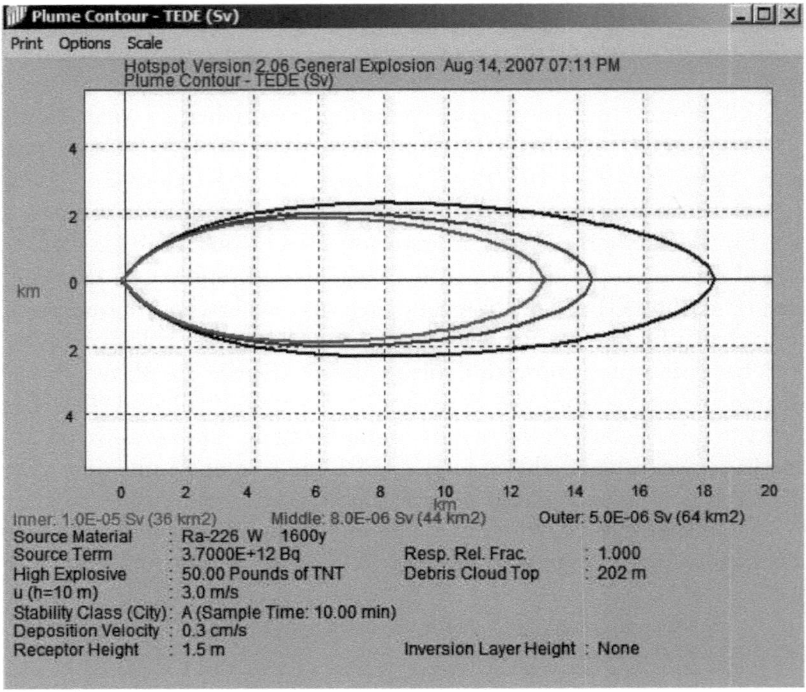

Figure 7.4 Radium-based *dirty bomb* TEDE-contour plot in an urban environment (^{226}Ra activity: 3.7 TBq).

(i.e., the health risk associated with ingestion is exceeded significantly by its health effects as a heavy metal, causing kidney toxicity rather than radiation-induced cancer).

7.5.5.2. Radium-based dirty bomb

Radium (^{226}Ra) makes a more effective *dirty bomb* in terms of radiological impact than a uranium-based device because of its higher specific activity. Figure 7.4 shows the result of detonating a device with the same characteristics in an urban environment; due to building effects on atmospheric conditions the plume is significantly broadened.

Once the radionuclide is released due to the explosion, its further dispersion is determined mainly by meteorological conditions and surface roughness (i.e., there will be significant variation reflecting the environmental conditions at the time of the uncontrolled release). As an example, radiation dose iso-curves for the first 4 h after the detonation of a radium-based dirty bomb (^{226}Ra activity: 370 GBq; explosives: about 50 kg TNT) are shown in Figure 7.5.

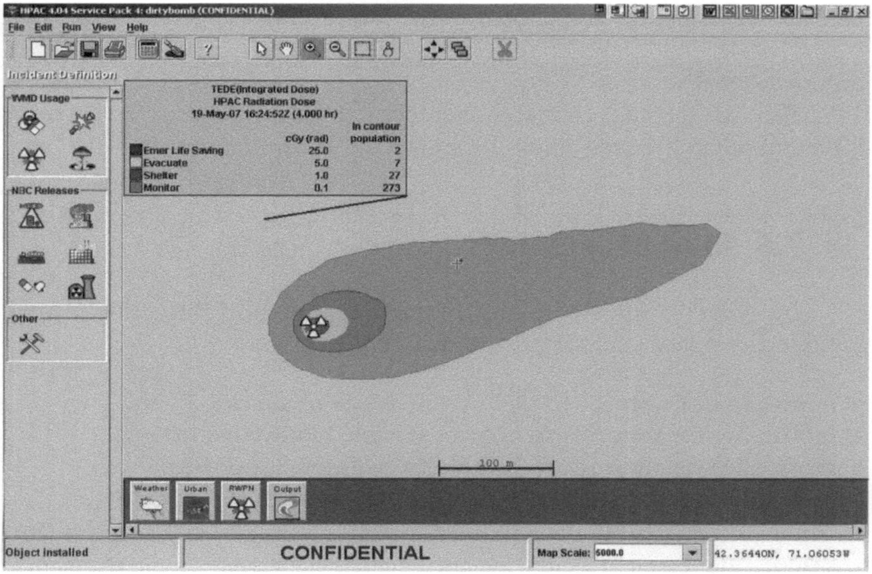

Figure 7.5 Dose iso-curves for the first 4 h after the detonation of a radium-based dirty bomb (^{226}Ra activity: 370 GBq; explosives: about 50 kg TNT).

7.5.6. Radiological exposure device (REXD)

A radiological exposure device exposes people in the vicinity of the device to radiation. A possible threat scenario foresees the covert irradiation of one or more individuals by placing a strong source in a high traffic density area, exposing a large number of people in a short time. This could result in a high individual external dose.

Alternatively, such covert irradiation of a group of persons could be achieved by placing a weaker source in a high traffic density area, exposing a limited number of people over a longer period of time. In any case, this mode of terror attack would result in a large collective dose but low individual external dose.

A variation of the REXD based on internal exposure from ingestion or inhalation received much attention last year through the previously discussed case of the fatal poisoning of former Russian spy Alexander Litvinenko with the alpha-emitting radionuclide polonium (^{210}Po). The fatal ^{210}Po poisoning of Alexander Litvinenko apparently occurred by ingestion. The actual date of the poisoning is not clear; hence, there is some uncertainty in the estimates of the amount of ^{210}Po and doses calculated. The estimated intake based on available post-mortem data is 15 GBq or in terms of mass about 0.09 mg of ^{210}Po. One expert estimate of the total organ doses from a soluble form of polonium in the body for 22 days resulted in very high doses to the red bone marrow of 130 Gy, 322 Gy to the liver,

583 Gy to the kidney, and 532 Gy to the spleen (http://www.wmdinsights.com/I18/I18_R1_LitVinenkoPoisoning.htm, *Mark L. Maiello, Ph.D*). A minimal lethal dose of ^{210}Po for an 80-kg person has been estimated as 148 MBq or 0.89 μg (www.answers.com/topic/polonium). Apparently, Litvinenko had ingested about 100 times the lethal amount of ^{210}Po.

7.6. Risk Assessment

In order to assess the risk of radiological terrorism using natural radionuclides, it is necessary to determine the:

- motivation of terrorist to deploy this mode of attack;
- probability for terrorists to acquire suitable radioactive material;
- probability for terrorists to be able to build such a device;
- probability for terrorists to be able to deploy such a device;
- consequences on population under attack and environmental contamination; and
- capability of society to implement effective countermeasures enabling it to manage the aftermath of such an attack.

Motivation: Terrorists have already indicated several times that they are interested in threatening society with uncontrolled exposure to radiation sources, albeit not to natural radioactivity. For example, in 1995, the Chechen field commander Shamil Basaev informed the Russian media about a radiation source he had ordered buried in a Moscow park – statement for the record by Dr. Gordon C. Oehler, director, Nonproliferation Center, to the Senate Armed Services Committee, "The Continuing Threat From Weapons of Mass Destruction, Appendix A: Chronology of Nuclear Smuggling Incidents" March 27, 1996. However, the only proven case where a natural radioactive nuclide was deliberately selected for the criminal act was the murder of Alexander Litvinenko in November 2006 with ^{210}Po.

Logistical capability: Based on the data contained in the DSTO database, ^{226}Ra thefts and seizures account for only 6% (7 cases) of all security-relevant cases. The low absolute number of incidents indicates that the main interest of criminals lies with man-made radionuclides as the material of choice.

Technical and operational capability: In Chechnya, terrorists stole radioactive waste from the RADON storage facility near Grozny and combined the man-made radioactive material with conventional explosives. Such an operational device was discovered by the Russian forces next to a railway line, attached to a mine buried underground – "Container with Radioactive Substances Found in Chechnya," ITAR-TASS, December

29, 1998 (in DSTO). Also, criminals have already used strong radiation sources to covertly kill their opponents (Ward, 1994; see also Korolkov, 2005). Therefore, it should not pose an insurmountable problem to obtain suitable naturally radioactive material and construct a functioning device, if they wished to do so.

Consequences: Dispersal of radioactive material with a dirty bomb does not require a large amount of explosives. Therefore, the physical damage by a dirty bomb with a low explosive yield will be small. In the immediate vicinity of such an explosive release of radioactive dust, people are likely to be injured or killed by the explosion itself, and contaminated with radioactive debris. Beyond a few tens of meters, only the radioactivity will be of concern. The associated health effects are due to the product of dose rate and exposure time. Typically, most of the radioactivity of concern will be in the form of fine dust and will settle out within hundreds of meters of the explosion at most. A small fraction of the respirable dust will be carried to tens of kilometers, depending on the physical state of the radioactive material, the wind, the roughness of the terrain, and whether there is a temperature inversion.

In summary, survivors of a dirty bomb attack in the immediate vicinity of the explosion will in all likelihood also survive the radiological impact. In an urban environment, the blast could result typically in the destruction of several cars and possibly set a few buildings on fire. The number of dead and injured would not be any different from a similar detonation without any radioactive material involved.

The intentional contamination of foodstuff can result in excessive individual doses, ultimately leading to an increased cancer risk or even death.

In the case of a covert irradiation of a large number of persons with a strong radiation source or the dispersion of radioactive aerosols, this would result in at most a large collective dose, while high individual doses are unlikely, depending on the source strength, duration of exposure, and distance between source and the irradiated persons.

Effective countermeasures: Irrespective of the mode of attack, the disturbance in terms of psychological impact would strain the resources of society in order to undo the consequences of the radiological attack. In the case of a dirty bomb attack, an additional major problem would be the significant economic loss, since it is a major economic weapon (Steinhäusler and Edwards, 2005).

Table 7.3 summarizes the qualitative assessment of the different components in order to determine the resulting risk for the various modes of radiological terrorism using natural radionuclides.

The risk of terrorists using radioactive aerosols is relatively low, since it requires a certain level of sophistication to generate respirable aerosols of the optimum size distribution. Also, the detrimental consequences for the target area are rather limited to surface contamination which can be removed by various techniques.

Table 7.3 Qualitative risk assessment for the various modes of radiological terrorism using natural radionuclides.

Parameter	Radioactive solution	Radioactive aerosol	Dirty bomb	Radiological exposure device
Motivation	Low	Low	Low	Medium
Acquisition of material	Medium	Medium	High	Medium
Technical/operational capability to build/deploy	High	Low	High	High
Consequences – man	Low	Low	Medium	Medium
Consequences – environment	Medium	Low	High	Low
Effective countermeasures	Medium	High	Medium	Low
Risk	Medium	Low	High	Medium

The deployment of radioactive solutions (e.g., covert spraying streets or the intentional contamination of drinking water systems) is within the technical and operational realm of capabilities of international terrorist organizations. The potentially significant disturbance of society and the limited range of possible countermeasures result in an overall medium risk. This also applies to radiological exposure devices: although the degree of motivation is obviously higher and terrorists undoubtedly have the capability to master such an attack, the consequences are limited to a few individuals rather than a large segment of society. Therefore the overall risk is rated as medium.

The highest risk for society is associated with the deployment of a dirty bomb. First of all, it has been indicated repeatedly that terrorists consider it a potential weapon of choice. Secondly, they have the capability to build and deploy such a device once they have acquired suitable radioactive material. Thirdly, consequences to society can be extraordinarily detrimental (high cleanup costs, lasting property devaluation, prolonged loss of business, stigmatization of victims), which raises the attraction to terrorists.

7.7. Societal Response

7.7.1. Professional experience

Currently, society has had little experience with terrorism deploying natural radioactivity. However, the performance of the international community in response to the only case hitherto was rather disappointing, indicating that it is inadequately prepared to manage even a single radiological terror attack, involving one victim only. It is worthwhile to look again at the case of Alexander Litvinenko and the practical health physics problems it entailed in 2007:

- there was no pre-established *protective action guide* for people;
- there was no well-established background level for ^{210}Po in urine; and
- there was no cytogenetic or other biomarker analysis which could distinguish whether it was low or high LET radiation that delivered the deadly damage.

These deficits, coupled with a lack of initial sharing information by the UK Health Protection Agency (HPA), produced an information void in Europe and the United States:

- no data on the concentration found in contaminated persons;
- no data on environmental contamination at various locations in the United Kingdom; and
- despite a lack of information on several essential health physics criteria, HPA publicized statements such as *the event was nothing of concern.*

Several other countries would be faced with similar challenges if terrorist activity succeeds in spreading radioactive material in a public setting (Whitcomb, 2007). The significant social impact of low-level irradiation (doses of 0.7 Gy or less) has already been acknowledged. While this radiation dose level is not likely to result in a performance decrement, the perception of having been exposed, however, can result in severe psychological casualties. Psychosomatic symptoms in persons exposed to radiation can cover a wide range, such as fear, stress, depression, neurasthenia, hypochondria, deficits in memory and attention, as well as long-term fear of radiation.

A major reduction in the cost and harm of an RDD incident would come from action by the competent authorities to provide a more realistic balance of benefits versus risks from radiation. This would entail establishing a level for exemption, clearance, or authorized release for certain areas and incidents, compatible with official survey and labeling of the average exposure levels expected during the emergency and recovery period.

Typical background radiation experienced by the public is some 3 mSv y^{-1}, of which half comes from diagnostic medical or dental procedures – a factor of 300 above a hypothetical criterion of 10 μSv y^{-1}, frequently defined as "negligible radiation." For a uniform distribution of a given amount of dirty dust, choosing the 10 μSv y^{-1} criterion can result in the forced evacuation of 300 times the area as would be required at the 3 mSv y^{-1} level.

A criterion that might be used by authorities and individuals is to compare an inferred life value of $1–5 million with the cost of relocation, which might be $20,000 per person. A person valuing his or her life at $5 million might decide to relocate at the cost of $20,000 if by so doing a probability of death of $20,000/$5 million = 0.4% could be avoided.

At 0.05 probability of death per sievert of exposure, 0.4% would correspond to 80 mSv, or some 27 years of $3\,\text{mSv}\,\text{y}^{-1}$ background radiation.

7.7.2. Local authorities and self-help

News moves fast and informally in the modern information-rich society. If the population is not to panic from news of a "radiation attack," the local authorities must have worked together with the competent central authority and expert consultants to issue a previously prepared communication, within about 5 min, that contains an early assessment and initial instructions on what to do. Therefore, practically applicable information should be made available immediately to all who are interested in such a "feed" and the local inhabitants alerted. Such instructions and auxiliary information must also be provided on the web in a form readily and permanently accessible and one that will not be overwhelmed by the ensuing queries. The direction of the plume of dirty dust should be part of the initial report.

First responders and health professionals are dedicated to saving lives and caring for the wounded or exposed, but they must be protected from injury themselves. First responders must be empowered with knowledge and support; it should not be taken for granted that they will all accept to respond to such events. It is critically important to maintain their motivation to volunteer for such service.

Members of the public at some distance from the event should ordinarily remain indoors where they are until the situation is assessed, while those who are contaminated with radioactive materials must help themselves to reduce further exposure. Much of this decontamination would have to be handled by members of the public themselves; as they enter buildings for shelter from the passing cloud of dirty dust, they should remove their shoes and outer clothing. Buildings with heating, ventilating, and air-conditioning systems should be able to close their air inlets until the cloud of radioactivity passes by, so as to reduce radioactive contamination inside the buildings and to accommodate "sheltering in place" until the plume of radioactivity has passed.

7.7.3. Integrated medical response

Contaminated individuals should not be taken to hospitals, which might as a result become so contaminated that their ordinary use would be impaired. Instead, improvised facilities should be planned, including the provision of germ-free rooms or tents that could be provided in facilities with open interior space. Although few people may actually require urgent care in an RDD event, a very large number might demand admission to normal urgent care facilities, not only overwhelming but also contaminating them so that they could no longer be used.

Radiological terrorism thus presents a unique set of global medical challenges and the necessity for public self-help. Among these challenges are ensuring public situation awareness, timely casualty triage, and expeditious acute and definitive treatment. The policy recommendations include field expedient placement and management of casualties in distributed contingency facilities using specially trained medical personnel in the context of leveraging existing medical infrastructures, including public health. "Contingency" facilities may include ships, warehouses, and sports stadia, and other non-traditional facilities such as mobile medical modules, and should be accompanied by "lockout" of hospital facilities to radiological injuries. Lockout will require enhanced physical security.

Mass casualty management requires the use of expedient facilities such as tents that will also accommodate stabilization of combined injury patients such as surgical patients. This will require an integrated response plan in which medical resource and treatment is central to logistical support and allocation of resources such as communication bandwidth and supplies. Rather than a gradual approach to medical resource buildup, urgent medical care should be particularized for more expedient logistics and care at the point of injury.

In the special case of dirty dust incidents in the urban underground transport system, the medical management should occur proximate to the site of injury with cooperation of rail personnel to expedite transport of casualties. Respiratory protection is essential for prevention of continued inhalation exposure and use of a small blower providing locally filtered air is recommended for response personnel and injured patients.

OVERVIEW

The introduction (Chapter 1) of the book starts with the question – Is *technological enhanced natural radiation* (TENR) a Universal issue? The reply to this question is a qualified positive, because the neutrino interacts very little with matter, and from the radiological point of view it does not need to be taken into account. However, TENR is indeed considered to be a global issue. Although radioactive emissions are in the lower end of the Maxwell electromagnetic spectrum, natural radiation *per se* includes the whole spectrum (i.e., 10^{-15} m $\leq \lambda \leq 10^4$ m). An overview of the human activities that modify NORM into TENORM as products, byproducts, or residues is presented at the end of the introduction. Individuals may receive increased radiation exposure due to TENR either as workers or consumers. In addition, environmental conditions may change as a result of several NORM/TENORM/TENR-related industrial processes. This overview shows flowcharts representative of industrial processes in which NORM, TENORM, or TENR have environmental implications.

The second chapter discusses the fact that DU, which has been used militarily more often recently, has provoked assessments of its potential health and environmental impacts. There is an enormous inventory of DU stocks throughout the world ($\approx$1.2 million tons), of which the United States and Russia are responsible for about 80%, divided almost equally between the two. The WHO examined the DU problem and reached a number of conclusions, which include: significant impact of DU can be achieved only by military use; young children playing within a former conflict zone could receive a DU exposure due to ingestion of contaminated soil because of hand-to-mouth activity; and there are gaps of knowledge and thus further research is recommended in some areas to improve health risk assessment. The Institute of Medicine (IOM) of the United States National Academies has also examined the DU problem and made a series of recommendations which are part of a 2008 document entitled "Epidemiological Studies of Veterans Exposed to Depleted Uranium: Feasibility and Design Issues," published by the National Academy Press.

In this book, the third chapter is the most lengthy. It discusses, to the extent compatible with the scope of the book, the terrestrial, atmospheric, and aquatic natural radioactivity as far as NORM/TENORM/TENR industries are concerned. Considering that there are 340 nuclides in nature, from which 70 are radioactive, the discussion restricts itself to the ^{235}U, ^{238}U, and ^{232}Th radioactive series plus ^{40}K characteristics. The

Radioactivity in the Environment, Volume 17
ISSN 1569-4860, DOI 10.1016/S1569-4860(09)01708-2

misinterpretation of the contribution of muons to the external radiation measured with common radiation detectors at sea level is pointed out, as well as its implication on the dose coefficients usually accepted by the radioprotection community for external gamma radiation. A comparison is made of the dose coefficients for external gamma radiation coming from the soil as estimated at different times by several investigators, and as recommended by ICRP and UNSCEAR. The differences in geology throughout the world are significantly less important, as far as the contributions to external gamma dose from soils are concerned, than those related to the enhancement of natural radionuclides in soils due to anthropogenic industrial activities. The contribution of the primordial radionuclide ^{87}Rb to human exposure is briefly examined just to conclude that it is of little importance. Other non-chain naturally occurring primordial radionuclides are listed, but their contributions from the radiological point of view are even less important than that of ^{87}Rb. The fact that the averaged radon flux adopted by UNSCEAR is about two times higher than those values measured or estimated by investigators in several parts of the world suggests that it is necessary to revise either the parameters used or the methodology to average the worldwide estimates. Oil and gas, coal, mineral sands, phosphate fertilizer, and the recycling industries are examined as far as their NORM, TENORM, and TENR implications are concerned. The main NORM problem in the oil industry is the wide range of ^{226}Ra and ^{228}Ra activity concentrations found in scales and sludges resulting from E&P operations. The authors suggest a concerted international effort to deal with the diffuse NORM problem of the oil and gas industry. The typical NORM problem is sometimes aggravated by misinterpretations of the earlier versions of the London Convention which prohibited, *inter alia*, dumping high-level radioactive wastes into the seas and oceans, but allow low-level radioactive wastes to be dumped under special permits. The biological availability of some selected NORM radionuclides makes it reasonable to hypothesize that the alpha radiation dose resulting from high bioaccumulation factors may be considerable. Current efforts by the International Atomic Energy Agency (IAEA) and the European Community (EC), for example, are deemed as urgent and necessary. The magnitude of the NORM problem in the coal industry is described. A comparison that was made in the past between the radioactivity releases from coal-fired and nuclear power plants was quite favorable to the latter. Because the total volume of wastes from the coal industry can amount to hundred of thousands of cubic meters, such wastes may represent a significant radioecological risk. Dose estimates from TENORM typical of the coal industry are presented. A brief history of the mineral sands industry allows the reader to understand the origins of some orphan sources related to this industry. The amount of TENORM wastes generated in the monazite cycle is quite significant – for example, for each

100 mt of monazite chemically processed, approximately 2% of cake II (i.e., Th and U oxides) and between 6 and 7% mesothorium (Ba, Ra SO_4) are produced. The typical TENORM remnants left behind contain activity concentrations (mostly ^{226}Ra plus ^{228}Ra combined) as high as 10^4 kBq kg^{-1}. Attempts have been made to solve problems associated with TENORM residues of the monazite cycle, but thus far to no avail. The titanium pigment and the gas mantle industries are also examined in connection with their occupational and TENORM wastes problems. The three-phase plan for the NORM/TENORM industries put forth by the Council Directive of 96/29/EURATOM in 1996 is discussed, but it is also pointed out that the development of the proposed three phases may be complicated by the fact that disposals of residues of the NORM/TENORM industries are sometimes made into conventional waste repositories. The phosphate industry has been under review by the "Institut Mondial du Phosphate" to evaluate the impact of the Directive 96/29 on its operations and residues. The recycling industry for scrap metals is also examined, but it is concluded that, with proper control of metal scrap processing (i.e., active enforcement of international regulations with fines for violators), this industry does not pose a problem of major concern from the radiological viewpoint.

In Chapter 4, it is explained that human and other living beings are fairly well shielded from cosmic radiation at sea level, but the external exposure due to muons is still to be better understood. Populations living at high altitudes, astronauts, military fighter pilots, and commercial jet airplane travelers (like crews and frequent travelers) are the segments of the human population, which are most likely to receive enhanced exposures to cosmic radiation. The chapter presents fairly well-based estimates and measurements reported for the doses to each of those segments. The effects of cosmic radiation on avionics are also examined.

Chapter 5 presents an account of the relationship of the prehistory of the metrology of ionizing radiation and the use of alpha particles emitted by selected naturally occurring radionuclides as tools to measure quantities like electric charge per mass and particle velocity. Historical aspects of radon measurements and units, the metrology before WWII and its improvements during the Manhattan Project are also examined. The need for an internationally accepted NORM metrology began in the 1990s and triggered national and international intercomparison exercises. In the foreseeable future, metrology will face new challenges to cope with the applications of NORM/TENORM/TENR regulations in developed, as well as developing, countries. Environmental modeling based on NORM/TENORM contamination is usually represented by time-dependent (or -independent) differential equation compartments as follows: source term → transport and interaction terms → concentration or dose. The ongoing IAEA EMRAS program, already in its second round, has recognized the importance of the NORM issue.

Chapter 6 deals with legal aspects of the NORM/TENORM/TENR issue. Although it is well recognized that hundreds of millions of tons of NORM/TENORM wastes are generated every year from a wide range of activities – like contaminated ion exchange resins from drinking water treatment, ash from coal-fired power plants, and contaminated water and tailings from a myriad of extracting industries – the international approach to deal with the NORM/TENORM issue is still heterogeneous. Many countries lack specific regulations limiting NORM/TENORM exposures. The leadership of the IAEA is certain to establish a comprehensive guidance on NORM/TENORM radiation protection for workers and the public in general. A table is presented in this chapter with recommended values for limiting radiation exposure for occupational and public exposure. Article 42 of Directive 96/29 regulates the radiation exposure of aircraft crews.

Chapter 7 discusses the use of natural radionuclides as a terrorist weapon, though some technical details are omitted for reasons of security. A subchapter examines how suitable some natural radionuclides are to be used with criminal intent and for radiological terror attacks. Illegal acquisition and illicit trafficking of natural radionuclides to be used in terror attacks are examined, as well as the potential motivations for such attacks. Some important aspects of selected effects of a radiological dispersive device (RDD) and a radiation emitting device (RED) are presented. Generic scenarios involving contamination of drinking water and roads are discussed. The uses of contaminated aerosols in aircraft and indoors are discussed, as well as the so-called dirty bomb (i.e., a RDD that disperses radioactivity by means of a conventional explosive propellent). Risk assessments of radiological terrorism, societal, and medical responses based on known experiences are offered to the reader and to the authorities responsible for such responses.

A comprehensive set of references is included to support the statements made in the book.

REFERENCES

Aarkrog, A. (1977). Environmental behavior of plutonium accidentally released at Thule, Greenland. *Health Physics*, 32, 171–173.

Aarkrog, A., P. Ästrup, G. Asmund, P. Bjerregaard, D. Boertmann, L. Carlsen, J. Christensen, M. Cleemann, R. Dietz, A. Fromberg, E. Storr-Hansen, N. Z. Heidam, P. Johansen, H. Larsen, G. B. Paulsen, H. Petersen, K. Pilegaard, M. E. Poulsen, G. Pritzl, F. Riget, H. Skov, H. Spliid, P. Weihe and P. Wåhlin. (1997). AMAP Greenland 1994–1996. Arctic Monitoring and Assessment Programme. Danish Environmental Protection Agency, Environmental Project No. 356: 788 pp.

Adams, L., E. J. Daly, R. Harboc-Sorensen, R. Nickson, J. Haines, W. Schafer, M. Conrad, H. Griech, J. Merkel, T. Schwall, and R. Heaneck. (1992). A verified proton induced latch-up in space. *IEEE Transactions on Nuclear Science*, 39, 1804–1808.

Advisory Committee for Radiation Biology Aspects of the SST (ACRBA). (1975). Final Report. Cosmic radiation exposure in supersonic and subsonic flights. *Aviation, Space Environmental Medicine*, 46, 1170–1185.

Ahrens, L. (1940). Measuring geologic time by the strontium method. In: *Geochronology: Radiometric Dating of Rockes and Minerals* (Ed. C. T. Harper). Douden, Hutchingson & Ross, Inc., Stroudsburg, PA, (1973), pp. 224–242 – Published originally in *Bulletin of the Geological Society of America*, 60, 217–266.

Äkerblom, G. (1986). Investigation and mapping of radon risk areas. Paper presented at the International Symposium on Geological Mapping in the Service of Environmental Planning Commission for the Geological Map of the World, Trondheim, Norway, 6–9 May, 1986. In: *Geology for Environmental Planning* (Ed. F. C. Wolff), Geological Survey of Norway, Special Papers, pp. 96–106.

Alan, M. N., M. I. Chowdhury, M. Kamal, S. Ghose, M. Islam, M. N. Mustafa, M. M. H. Miah, and M. M. Ansary. (1999). The ^{226}Ra, ^{232}Th and ^{40}K activities in beach sand minerals and beach soils of Cox's Bazar, Bangladesh. *Journal of Environmental Radioactivity*, 46, 243–250.

Allisy, A. (1994/1995). From Curie to the Becquerel. *Metrologia*, 31, 467–479.

Al-Masri, M. S., and A. Aba. (2007). First proficiency test for the determination of NORM in contaminated soil from the oil field. *Accreditation and Quality Assurance: Journal for Quality, Comparability and Reliability in Chemical Measurement*, 12, 249–256. 10.10.1007/s00769-006-0246-1

Al-Masri, M. S., and R. Shwiekani. (2008). Radon gas distribution in natural gas processing facilities and workplace air environment. *Journal of Environmental Radioactivity*, 99, 574–580.

Alpher, R. A., and R. C. Herman. (1948). On the relative abundance of elements. *Physical Review*, 74, 1737–1742.

American Petroleum Institute (API). (1988). *NORM Regulatory Analysis Report*. Task Group on Regulatory Analysis. American Petroleum Institute, Washington, D.C.

American Petroleum Institute (API). (1992). *Bulletin on Management of Naturally Occurring Radioactive Materials (NORM) in Oil and Gas Production*. API Bulletin E2. American Petroleum Institute, Washington, D.C.

American Petroleum Institute (API). (1996). *A Naturally Occurring Radioactive Material (NORM)*. Disposal Cost Study, 1st Ed., API Publication 7100, API, Washington D.C.

Anderson, C. C., and H. H. Hinson. (1951). *Helium-bearing natural gases of the United States, analyses and analytical methods*, US Bureau of Mines Bulletin 486, 141pp.

Anjos, J. C., A. F. Barbosa, A. Bernstein, N. S. Bowden, W. Fulgione, E. Kemp, J. Magnin, H. Nunokawa, O. L. G. Peres, D. Reyna, A. Schilithz, R. C. Shellard, and R. Zukanovich Funchal. (2006). Angra dos Reis reactor neutrino oscillation experiment. *Brazilian Journal of Physics*, 36, 1118–1123.

Armstrong, T. W., R. G. Alsmiller, and J. Barish. (1969). Calculation of the radiation hazard at supersonic aircraft altitudes produced by an energetic solar flare. *Nuclear Science Engineering*, 37, 337–342.

Armstrong, T. W., and K. C. Chandler. (1972). HETC – A high energy transport code. *Nuclear Science and Engineering*, 49, 110.

Arnold, J. R., and E. C. Anderson. (1957). The distribution of Carbon-14 in Nature. *Tellus*, 9, 28–32.

Arnold, J. R., and W. F. Libby. (1949). Age determinations by radiocarbon content: Checks with samples of known Age. *Science*, 110, 678–680. 10.1126/science.110.2869.678.

Ashkenazi, R., J. Koch, and I. Orion. (2005). *Theoretical Assessment of Aircrew Exposure to Galactic Cosmic Radiation Using the FLUKA Monte Carlo Code*. Report from Soreq Nuclear Research Centre, Israel. Presented at the 22th Conference of the Nuclear Societies, 17–18 February 2004, Israel.

Avadhami, D. N., H. M. Mahesh, N. Karunakara, Y. Narayana, H. M. Somashekarappa, and K. Sidappa. (2005). Distribution and behavior of natural radionuclides in soil samples of Goa on the southwest coast of India. In: *The Natural Radiation Environment VII. Proceedings of an International Symposium*, Rhodes, Greece, 20–24, May 2002 (Eds J. P. McLaughlin, S. E. Simopoulus, and F. Steinhäusler), Elsevier, Amsterdam, pp. 1131–1140.

Baker, D. N. (1994). Mission investigates radiation arriving at Earth. *Eos*, 75, 130–131.

Baker, D. N., S. G. Kanekai, X. Li, S. P. Monk, J. Goldstein, and J. L. Burch. (2004). An extreme distortion of the Van Allen belt arising from 'Hallowe'en' solar storm in 2003. *Nature*, 432, 878–881. 10.1038/nature03116.

Bao, X. (2008). Uranium Solubility in Planetary Cores: Evidence from high Pressure and temperature experiments. In: Neutrino Geoscience 2008. 17–19 September 2008, Sudbury, Ontario, Canada.

Barashenkov, V. S., N. M. Sobolevskii, and V. D. Toneev. (1972). The penetration of beams of high-energy particles through thick layers of material. *Atomnaya Energiya*, 32, 217. In: Geological Aspects of Radon Mapping, XXX.

Bartlett, D. T. (1999a). Aspects of the exposure of aircraft crew to radiation. *Radiation Protection Dosimetry*, 81, 243–245.

Bartlett, D. T. (1999b). Radiation protection concepts and quantities for the occupational exposure to cosmic radiation. *Radiation Protection Dosimetry*, 86, 263–268.

Bartlett, D. T. (2004). Radiation protection aspects of the cosmic radiation exposure of aircraft crew. *Radiation Protection Dosimetry*, 4, 349–355.

Bäshounek, F., V. Santholzer, and F. Ulrich. (1935). Radioactivity in oil-waters in Czechoslovakia. *Nature*, 136, 910–911.

Baxter, M. S. (1993). Environmental radioactivity: A perspective on industrial contributions. *International Atomic Energy Agency (IAEA) Bulletin*, 2, 33–38.

Baxter, M. S. (1996). Technologically enhanced radioactivity: An overview. *Journal of Environmental Radioactivity*, 32, 3–17.

Baxter, M. S., S. Ballestra, J. Gastaud, T. F. Hamilton, I. Harms, L. Huynh-Ngoc, L. Liong Wee-Kwong, P. Parisi, H. P. L. Pettersson, P. P. Povinec, and A. Sanchez. (1996). Marine radioactivity studies in the vicinity of sites with potential radionuclide release. In: *Environmental Impact of Radioactive Releases. Proceedings of an International Symposium*,

International Atomic Energy Agency (IAEA), Vienna, Austria, 8–12 May 1995, IAEA Vienna, pp. 125–141.

Beck, H. L. (1972). The physics of environmental gamma radiation fields. In: *The Natural Radiation Environment II* (Eds J. A. S. Adams, W. M. Lowder, and T. F. Gesell), US Energy Research and Development Administration (USERDA) Report CONF-720805-P2, pp. 101–134.

Beck, H. L. (1980). *Factors for radionuclides deposited on the ground*. EML Environment Measurements Laboratory, Report 378, US Department of Energy, New York.

Beck, H. L., and W. M. Lowder. (1966). Cosmic ray ionization in the lower atmosphere. *Journal of Geophysical Research*, 71, 4661–4668.

Becquerel, A. H. (1896). Sur les radiations émises par phosphorescence. *Comptes Rendus Hebdomadaires des Seances de l'Academie des Sciences Paris*, 122, 421–422.

Bedrosian, P. A., D. G. Easterly, and S. L. Cummings, Radiological Survey around Power Plants using Fossil Fuels. Easterns Environmental Radiation Laboratory – United States Environmental Protection Agency (USEPA) – EERL 71–73.

Beer, J. (2000). Long-term indirect indices of solar variability. *Space Science Reviews*, 94, 53–66.

Bell, K. G., C. Goodman, and W. L. Whitehead. (1940). Radioactivity of sedimentary rocks and associated petroleum. *Bulletin of American Association of Petroleum Geologists – (AAPG) Bulletin*, 24, 1529–1547. 10.1306/3D933230-1681-11D7-8645000102C1865D.

Bella, F. and E. Pettinelli. (1989). Radon monitoring aimed to study seismic precursors. In: *Radon Monitoring in Radioprotection, Environmental Radioactivity and Earth Sciences. Proceedings of an International Workshop, International Centre for Theoretical Physics (ICTP)*, Trieste, Italy, 3–14 April 1989 (Eds L. Tommasino, G. Furlan, H. A. Khan, and M. Monnin), World Scientific, Singapore, pp. 275–294).

Betti, M. (2003). Civil use of depleted uranium. *Journal of Environmental Radioactivity*, 64, 113–119. 10.1016/S0265-931X(02)00042-5.

Bikit, I., J. Slivka, M. Krmar, M. Vestović, L. Čonkić, E. Varga, S. Čurčić, and D. Mrdja. (2001). Determination of depleted uranium at the Novi Sad low-level laboratory. *Archive of Oncology*, 9, 241–243.

Bikit, I., J. Slivka, M. Krmar, M. Vestović, L Čonkić, E. Varga, S. Čurčić, and D. Mrdja. (2005). Depleted uranium determination in soil samples. In: *The Natural Radiation Environment VII. Proceedings of an International Symposium*, Rhodes, Greece, 20–24 May 2002 (Eds J. P. McLaughlin, S. E. Simopoulus, and F. Steinhäusler), Elsevier, Amsterdam, pp. 150–154.

Bikit, I., J. Slivka, D. Mrdja, N. Zikic-Todorovic, S. Čurčić, E. Varga, M. Vestović, and L. Čonkić. (2003). Simple method for depleted uranium determination. *Japanese Journal of Applied Physics*, 42, 5269–5273.

Bird, D. J., S. C. Corbato, H. Y. Dai, J. W. Elbert, K. D. Green, M. A. Huang, D. B. Kieda, S. Ko, C. G. Larsen, E. C. Loh, M. Z. Luo, M. H. Salamon, J. D. Smith, P. Sokolsky, P. Sommers, J. K. K. Tang, and S. B. Thomas. (1995). Detection of a cosmic ray with measured energy well beyond the expected spectral cutoff due to cosmic microwave radiation. *Astrophysical Journal*, 441, 144–163.

Blight, G. E., and J. A. Caldwell. (1984). The abatement of pollution from abandoned gold-residue dams. *Journal of the South African Institute of Mining and Metallurgy*, 84, 1–9.

Bochicchio, F., G. Campos Venuti, S. Manccioppi, S. Piermattei, S. Risica, L. Tommasino, and G. Torri. (1992). Natural radiation indoor exposure of Italian population. *Proceedings of the Eight International Congress of the International Radiation Protection Association (IRPA 8)*, Montreal, Canada, 17–22 May, Vol. 1, pp. 1561–1565.

Bochicchio, F., G. Campos Venuti, S. C. Nuccetelli, S. Piermattei, S. Risica, L. Tommasino, R. Tommasino, and G. Torri. (1994). The Italian survey as the basis of the national radon policy. *Proceedings of the First International Workshop on Indoor radon*

Remedial Action, Rimini, Italy, 27th June–2nd July 1993. Published in *Radiation Protection Dosimetry*, 56, 1–4.

Bohem, F., B. Cook, G. Gratta, H. Henrikson, J. Kornis, D. Lawrence, K. B. Lee, K. McKinny, L. Miller, V. Novikov, A. Piepke, B. Richie, D. Tracy, P. Vogel, Y.-F. Wang, and J. Wolf. (2001). Final results from the Palo Verde neutrino oscillations experiment. *Physical Review D*, 64, 112001-1–112001-10.

Bokoria, E., B. Michalik, L. Juhásza, G. Trabidouc, Iz. Chmielewskab. (2007). Fractionation of ^{226}Ra and ^{228}Ra in waste material of coal mine and fired power plant and the effect of speciation on plant uptake. In: *Naturally Occurring Radioactive Material (NORM V)*. Presented as Poster 3-3 at the NORM V International Symposium, Seville, Spain, 19–22 March. [Available in CD-ROM format – not included in the NORM V Proceedings.]

Boone Jr., W. J. (1958). *Helium-bearing natural gases of the United States, analyses and analytical methods*, Supplement to Bulletin 486. *Bureau of Mines Bulletin*, 576, 117pp.

Bottolier-Depois, J. F. (1997). *Evaluation de l'exposition au rayonnement cosmique a bord d'un avion long courier.* Institut de Protection et de Securite Nucleaire, IPSN Report SDOS/97-01.

Bottollier-Depois, J. F., Q. Chau, P. Bouisset, G. Kerlau, L. Plawinski, and L. Lebaron-Jacobs. (2003). Assessing exposure to cosmic radiation on board aircraft. *Advances in Space Research*, 32, 59–66.

Bottolier-Depois, J. F., Q. Quang Cau, P. Bouisset, G. Kerlau, L. Plawinski, and L. Lebaron-Jacobs. (2000). Assessing exposure to cosmic radiation during long-haul flights. *Radiation Research*, 153, 526–532.

Bouville, A., and W. M. Lowder. (1988). Human population exposure to cosmic radiation. *Radiation Protection Dosimetry*, 24, 293.

Bowden, N. S. (2008). Ractor monitoring and safeguards using antineutrino detectors. In: *Neutrino Physics and Astrophysics, Proceedings of the XXIII Conference* [*Journal of Physics: Conference Series*, 136, 1–6]. doi:10.1088/1742-6596/136/2/022008.

Bowen, V. T., V. E. Noshkin, and H. D. Livingston. (1980). Fallout radionuclides in the Pacific Ocean: Vertical and horizontal distributions, largely from GEOSECS stations. *Earth and Planetary Science Letters*, 49, 411–434.

Briquet, C. M., Cipriani, and A. C. Ferreira. (2004). Soil contamination in a rural site used for rare earth industry by-product disposal. In: *Naturally Occurring Radioactive Materials (NORM IV), Proceedings of an International Symposium*, IAEA-TECDOC-1472, International Atomic Energy Agency, Vienna.

Brooks, B. T. (1931). Chemical considerations regarding origin of petroleum. *American Association of Petroleum Geologists (AAPG) Bulletin*, 15, 611–615. 10.1306/30932984-16B1-1107-8645000102C1865D.

Brooks, B. T. (1948). The origin of petroleum in the light of recent research. *The Ohio Journal of Science*, XLVIII, 129–145.

Buchel, K. H., H. H. Moretto, and P. Wodfish. (2000). *Industrial Inorganic Chemistry*. Wiley-VCH, Weinheim.

Budinger, T. F., J. T. Lyman, and C. A. Tobias. (1972). Visual perception of accelerated nuclei interacting with the human retina. *Nature*, 239, 209–211. 10.1038/239209a0.

Bureau International de Poids et Measures (BIPM). (2005). What is metrology, http://www.bipm.org/en/bipm/metrology.

Burnett, R. J. (1998). Precious metals – gold. In: *Mineral Resources of Namibia*, http://www.gsn.gov.na/pdf/gold.pdf. Last accessed on 11 June 2009.

Burton, E. F. (1904). A radioactive gas from crude petroleum. *Philosophical Magazine*, 8, 498–508.

Cassette, Ph. (1996). The discovery of radioactivity. *Radioactivity & Radiochemistry*, 7, 3.

Cassete, P., M. Sahagia, L. Grigorescu, M. C. Lépy, and J. L. Picolo. (2006). Standardization of ^{226}Ra by LSC and comparison with α- and γ-spectrometry. *Applied Radiation and Isotopes*, 64, 1465–1470.

Castelfranchi, G. (1934). Fisica Moderna: Visione Sintetica della Fisica D'Oggi, Quarta Edizione Rifatta, Editore Ulrico Hoelpi, Milano, 699pp.

Castrén, O., A. Voutilainen, K. Winqist, and I. Mäkeläinen. (1985). Studies of high indoor radon areas in Finland. *The Science of Total Environment*, 45, 311.

Commission of European Communities (CEC). (1989). *The Radiological Exposure of the Population of the European Community from Radioactivity in North European Marine Waters-Project 'Marina'*. An International Seminar, 14–16 June, Bruges.

Commission of European Communities (CEC). (1990). *The Radiological Exposure of the Population of the European Community from Radioactivity in North European Marine waters-Project 'Marina'*. Commission of the European Communities, Luxembourg. Report No. EUR 12483.

Central Intelligence Agency (CIA). (2009). The world fact book. Rank order – natural gas production, https://www.cia.gov/library/publication/the-world-factbook/rankorder/2180rank.html. Accessed on 14 February 2009.

Chadwick, J. (1932a). Possible existence of a neutron. *Nature*, 129, 312.

Chadwick, J. (1932b). The existence of a neutron. *Proceedings of the Royal Society of London Series A*, 136, 692–708.

Chalupnik, S. (2005). Impact of radium-bearing mine waters on the natural environment. In: *The Natural Radiation Environment VII. Proceedings of an International Symposium* (Eds J. P. McLaughlin, S. E. Simopoulos, F. Steinhäusler), Radioactivity in the Environment, Elsevier Series (M. S. Baxter, Series Editor), Vol. 7, pp. 985–995, ISSN 1569-4860/doi:10.1016/S1569-4860(04)07110-4.

Chirkov, A. M. (1975). Radon as a possible criterion for predicting eruptions as observed at Karymsky volcano. *Bulletin of Volcanology*, 39, 126–131. 10.1007/BFO2596952.

Choncubhair, Ó. Nf., Mc Laughlin, J., and Tokonami, S. (2008). Passive measurements of thoron and its progeny in some dwellings in Ireland. In: *The Natural Radiation Environment VIII. Proceedings of an International Symposium*, Búzios, Rio de Janeiro, Brazil, 7–12 October 2007 (Eds Anselmo S. Paschoa, F. Steinhäusler), American Institute of Physics Conference Proceedings 1034, New York, pp. 16–19.

Clarke, R. H. (1995). ICRP recommendations applicable to the mining and minerals processing industries and to natural sources. *Health Physics*, 69, 454–460.

Clay, J. (1935). Results of the Dutch cosmic ray expedition 1933V. The variation of the radiation with atmospheric pressure. A barometric paradox. The influence of clouds. *Physica*, 2, 189–195.

Clucas, S. N., C. S. Dyer, and F. Lei. (2005). The radiation in the atmosphere during major solar particle events. *Advances in Space Research*, 36, 1657–1664, (Space Life Sciences: Aircraft and Space Radiation Environment).

Coles, D. G., J. W. T. Meadows, C. L. Lindeken. (1974). A convenient method for the direct measurement of uranium in soil samples using high-resolution Ge(Li) gamma-ray spectrometer. In: *Proceedings of the Conference on Nuclear Methods in Environmental Research*, University of Missouri-Columbia, Columbia, MO (Eds J. R. Vogt and W. Meyer) 7/29-31, CONF-740701.

Conen, F., and L. B. Robertson. (2002). Latitudinal distribution of Rn-220 flux from continents. *Tellus*, 54B, 127–133.

Cooper, M. B. (2005). *Naturally Occurring Radioactive Materials (NORM) in Australian industries – Review of Current Inventories and Future Generation.* A report prepared for the Radiation Heath and Safety Advisory Council. EnviroRad Services Ltd., ERS-006, 40pp.

Cosma, C., O. Cozar, T. Jurcut, C. Baciu, and I. Pop. (2005). Simultaneous measurement of radon and thoron exhalation rate from soil and building materials. In: *The Natural*

Radiation Environment VII. Proceedings of an International Symposium, Rhodes, Greece, 20–24 May 2002 (Eds J. P. McLaughlin, S. E. Simopoulus, and F. Steinhäusler), Elsevier, Amsterdam, pp. 699–705.

Coursey, B. M., B. Collé, and J. S. Coursey. (2002). Standards of radium-226 from Marie Curie to the International Committee for Radionuclide Metrology. In: *Radionuclide Metrology and Its Applications, Proceedings of an International Conference* (B. M. Coursey, H. Janszen, L. R. Karam, and M. J. Woods, Guest Editors). Published in *Applied Radiation and Isotopes*, 56, 5–13.

Cowan, C. L., Jr., F. Reines, F. B. Harrison, H. W. Kruse, and A. D. McGuire. (1956). Detection of the free neutrino: A confirmation. *Science*, 124, 103.

Crokett, R. G. M., G. K. Gillmore, P. S. Phillips, A. R. Denman, and C. J. Groves-Kirby. (2005). Radon anomalies preceding earthquakes which occurred in UK, in summer and autumn 2002. *Science of the Total Environment*, 364, 138–148. 10: 1016/j. scitotenv.2005.08.003.

Curie, P. (1908). *Œuvres de Pierre Curie. Publiées par les soins de la Société française de physique.* Gauthier-Villars, Paris.

Curie, E. (1937). *Madame Curie. A Biography by Eve Curie.* Doubleday, Doran & Company, Inc., Garden City, New York, Translated by Vincent Sheean.

Curie, P., S. Curie, and G. Bémont. (1898). Sur une nouvelle substance fortement radio-active, contenue dans la pechblenbe. *Comptes Rendus Hebdomadaires des Seances de l'Academie des Sciences Paris*, 127, 1215–1217.

Das, N. K., R. K. Bhandari, D. Gosh, P. Sen, and B. Sinha. (2005). Anomalous fluctuation of radon, gamma dose and helium emanating from a thermal spring prior to an earthquake. *Current Science*, 89, 1399–1404.

Dawood, Y. H., and H. H. Abd El-Naby. (2007). Mineral chemistry of monazite from black sand deposits, northern Sinai, Egypt: A provenance perspective. *Mineralogical Magazine*, 71, 389–406. 10.1180/minerag.2007.4.389.

Debierne, M. A. (1899). Sur une nouvelle matiére radio-active. *Comptes Rendus Hebdomadaires des Seances de l'Academie des Sciences Paris*, 129, 593–595.

Debierne, M. A. (1903). Sur la production de la radioactivité induite par l'actinium. *Comptes Rendus Hebdomadaires des Seances de l'Academie des Sciences Paris*, 136, 671–673.

de Freitas, A. C., A. S. Alencar, C. R. Coutinho, and A. S. Paschoa. (2007). Radiation exposure at sea level measurement between Rio de Janeiro and the Antarctic Peninsula. *Nuclear Instruments and Methods for Physics Research A*, 580/1, 644–647. 10.1016/j.nima.2007.05.058.

Degens, E. T., F. Khoo, and W. Michaelis. (1977). Uranium anomaly in Black Sea sediments. *Nature*, 269, 566–569.

Degrange, J.-P., and S. Lepicard. (2004). Evaluation of occupational radiological exposures associated with fly ash from French coal power plants. In: *Proceedings of International Conference on Naturally Occurring Radioactive Materials (NORM IV)*, Szczyrk, Poland, 17–21 May, IAEA-TECDOC-1472, Vienna, pp. 230–239.

Delgado, J. U., J. Morel, and M. Etcheverry. (2002). Measurements of photon emission probabilities from the decay of ^{226}Ra and daughters. *Applied Radiation and Isotopes*, 56, 137–143.

Deme, S., I. Feher, I. Apathy, G. Reitz, and Yu. Akatov. (2005). Twenty years of TLD measurements on board space vehicles by the Hungarian "Pille" system. In: *The Natural Radiation Environment VII. Proceedings of an International Symposium*, Rhodes, Greece, 20–24 May (Eds J. P. McLaughlin, S. E. Simopoulus, and F. Steinhäusler) Elsevier Radioactivity in the Environment Series Vol. 7 (M. S. Baxter, Series Editor), Amsterdam, pp. 916–925, ISSN 1569-4860, doi:10.1016/S1569-4860(04)07110-4.

de Meijer, R. J., F. D. Smit, F. D. Brooks, R. W. Fearick, H. J. Wörtche, and F. Mantovani. (2006). EARTH: Towards Earth AntineutRino TomograpHy (EARTH). *Earth, Moon, and Planets*, 99, 193–206. 10.1007/s11038-006-9104-8.

Department of Veterans Affairs. (1995). *VA Fact Sheet – VA Programs for Persian Gulf Veterans*. June, Department of Veterans Affairs, Washington D.C.

Diehl, P. (1999). Depleted Uranium: A by-product of the Nuclear Chain. In: *Depleted Uranium: A Post-War Disaster*. Laka Foundation, Part 7, http://www.wise-uranium.org.

Domogatsky, G. V., V. I. Kopeikin, L. A. Mikaelyan, and V. V. Sinev. (2006). Baksan: Neutrino geophysics at Baksan: On searches for antineutrinos and radiogenic-heat sources in the interior of the Earth. *Physics of Atomic Nuclei*, 69, 43–50. 10.1134/S1063778806010078.

Dorn, F. E. (1900). Die von radioactiven Substanzen ausgesandte Emanation. *Abhandlungen der Naturforschenden Gesellschaft (Halle)*, 23, 1–15.

Dyer, C. S., F. Lei, C. N. Clucas, D. F. Smart, and M. A. Shea. (2003). Calculations on observations of solar particle enhancements to the radiation environment at aircraft altitudes. *Advances in Space Research*, 32, 81–93.

Dyer, C. S., and P. R. Truscott. (1999). Cosmic radiation effects on avionics. *Radiation Protection Dosimetry*, 86, 337–342.

Eckerman, K. F., and J. C. Ryman. (1993). *External Exposure to Radionulides in Air, Water, and Soil*. Federal Guidance Report No. 12. EPA 402-R-93-081.

Eisenbud, M. (1982). Conclusions derived from the 1975 conference on areas of high natural radioactivity. In: *Natural Radiation Environment* (Eds K. G. Vohra, U. C. Mishra, K. C. Pillai, and S. Sadasivan) (Bombay, India, January 19–23, 1981). Wiley Eastern Limited, New Delhi, pp. 21–25.

Eisenbud, M., and T. F. Gesell. (1997). *Environmental Radioactivity: From Natural, Industrial, and Military Sources*. Academic Press, San Diego, CA, 4th Ed., 655pp.

Eisenbud, M., and H. G. Petrow. (1964). Radioactivity in the atmospheric effluents of power plants that use fossil fuels. *Science*, 144, 288–289.

El Afifi, E. M., M. A. Hilal, M. F. Attallah, and S. A. El-Reefy. (2009). Characterization of phosphogypsum wastes associated with phosphoric acid and fertilizers production. *Journal of Environmental Radioactivity*, 100, 407–412.

Enamoto, S. (2006). Experimental study of geoneutrinos with KamLAND. *Earth, Moon, and Planets*, 99, 131–146.

Energy Information Administration (EIA). (2008). World crude oil production, 1960–2007, http://www.eia.doe.gov/aer/txt/ptb1105.html. Accessed on 14 February 2009.

European Atomic Energy Community (EURATOM). (1996). Council Directive 96/29/EURATOM of 13 May, Laying down the basic safety standards for protection of the health of workers and the general public against the danger arising from ionizing radiation. *Official Journal of European Community*, 19(L159), 29.

European Commission (EC). (1996a). Directorate-General Environment, Nuclear Safety and Civil Protection: Approaches for Regulating Management of Large Volumes of Waste Containing Natural Radionuclides in Enhanced Concentrations. Nuclear Science and Technology, EUR 16956, Luxembourg.

European Commission (EC). (1996b). Exposure of air crew to cosmic radiation: A report of EURADOS Working Group 11. *The Radiation Exposure and Monitoring of Air Crew*. Radiation Protection 85, EURADOS Report 1996-01.

European Commission (EC). (1997). Radiation Protection 88: Industrial Processes Concerning with Title VII of the European Union Basic Safety Standards Directive. 96/29/Euratom of 13 May 1996, EC Luxemburg.

European Commission (EC). (2002). Directorate-General Environment: Practical Use of the Concepts of Clearance and Exemption – Part II: Application of the Concepts of

Exemption and Clearance to Natural Radiation Sources. Radiation Protection Publication RP 122, Part II, Luxembourg.

European Commission (EC). (2004). General Directorate Energy and Transport: Effluent and Dose Control from European Union NORM Industries – Assessment of Current Situation and Proposal for a Harmonised Community Approach. Radiation Protection Publication RP 135, Vols. 1 and 2, Luxembourg.

European Union (EU). (1996). EU-Grudnorm, Richtlinien der Europäischen Gemeinschaft.

Failla, G. (1923). Ionization measurements. *American Journal of Roentgenology*, 10, 48–56, (Reproduced in *Health Physics*, 1980, 38, 889–897).

Farouk, M. A., and M. Al-Soraya. (1982). ^{226}Ra as a standard source for efficiency calibration of Ge(Li) detectors. *Nuclear Instruments Methods*, 200, 593.

Fasso, A., A. Ferrari, A., and P. R. Sala. (2001). Electron-photon transport in FLUKA. In: *Status, Proceedings of the Monte Carlo 2000 Conference*, Lisbon, October 23–26 2000 (Eds A. Kling, F. Barao, M. Nakagawa, L. Tavora, P. Vaz), Springer-Verlag, Berlin, pp. 159–164.

Feng, J. L., and J. Kumar. (2008). Dark-matter particles without weak-scale masses or weak interactions. *Physical Review Letters*, 101, 1301–1305. 10.1103/PhysRevLett.101.231301.

Fermi, E. (1934). *Zeitschrift fur Physik*, 88, 161.

Fermi, L. (1954). *Atoms in the family: My life with Enrico Fermi*. University of Chicago Press, Chicago.

Ferrari, A., and M. Pelliccioni. (2003). On the conversion coefficients for cosmic ray dosimetry. *Radiation Protection Dosimetry*, 104, 211–220.

Ferrari, A., M. Pelliccioni, and R. Villari. (2004). Evaluation of the influence of aircraft shielding on the aircrew exposure through an aircraft mathematical model. *Radiation Protection Dosimetry*, 108, 91–105.

Foelsche, T. (1961). *Radiation Exposure in Supersonic Transports*. NASA TN D-1383.

Foelsche, T., and E. H. Graul. (1962). Radiation exposure in supersonic transports. *Atompraxis*, 8, 365–380.

Foelsche, T., R. B. Mendell, J. W. Wilson, and R. R. Adams. (1974). Measured and calculated neutron spectra and dose equivalent rates at high altitudes: Relevance to SST Operations and Space Research, NASA TND-7715, National Aeronautics and Space Administration, Washington, D.C. The largest ever recorded solar flare – in terms of particle intensity-occurred in August 1972; the flare with the hardest energy recorded happened in February 1956 with calculated equivalent doses of 29 mSv at 20 km above ground.

Forbush, S. E. (1937). On the effects in cosmic ray intensity observed during the recent magnetic storms. *Physical Review*, 50, 1108–1109.

Fiorentini, G., M. Lissia, F. Mantovani, and R. Vannucci. (2005). Geo-neutrinos: A new probe of Earth's interior. *Earth and Planetary Science Letters*, 238, 235–247.

Fischer, A. G. and M. A. Arthur. (1977). Secular variations in the Pelagic realm. In: *Deep-Water Carbonate Environments* (Eds H. E. Cook and P. Enos), Society of Economic Paleontologists and Mineralogists (SEPM), Special Publication No. 25, pp. 19–50.

Fotakis, M., L. Tsikritzis, N. Tzimkas, N. Kolovos, and R. Tsikritzi. (2008). Naturally occurring radionuclides of ash produced by coal combustion. The case of the Kardia mine in Northern Greece. In: *The Natural Radiation Environment VIII. Proceedings of an International Symposium*, Búzios, Rio de Janeiro, Brazil, 7–12 October 2007 (Eds Anselmo S. Paschoa, and Friedrich Steinhäusler), American Institute of Physics Conference Proceedings 1034, New York, pp. 326–329.

Fowler, W. A. (1967). *Nuclear Astrophysics – Jayne Lectures for 1965*. American Philosophical Society, Philadelphia, PA, 109pp.

Friedberg, W., K. Copeland, F. E. Duke, and E. B. Darden, Jr. (2005a). Annual effective dose of ionizing radiation from natural sources: United States based airline pilots compared with non-flying residents of the United States. *Advances in Space Research*, 36, 1653–1656.

Friedberg, W., K. Copeland, F. E. Duke, K. O'Brien, E. B. Darden, Jr. (2005b). Health aspects of radiation exposure on a simulated mission to Mars. In: The Natural Radiation Environment VII. *Proceedings of an International Symposium*, Rhodes, Greece, 20–24 May 2002 (Eds J. P. McLaughlin, S. E. Simopoulus, and F. Steinhäusler), Elsevier Radioactivity in the Environment Series Vol. 7 (M. S. Baxter, Series Editor), Amsterdam, pp. 894–901, ISSN 1569-4860, doi:10.1016/S1569-4860(04)07110-4.

Friedberg, W., F. E. Duke, L. Snyder, D. N. Faulkner, K. O'Brien, E. B. Darden, Jr., and D. E. Parker. (1993). The cosmic radiation environment at air carrier flight altitudes and possible associated health risks. *Radiation Protection Dosimetry*, 48, 21–25.

Friedberg, W., D. N. Faulkner, L. Snyder, E. B. Darden, and K. O'Brien. (1989). Galactic cosmic radiation exposure and associated health risks for the air carrier crewmembers. *Aviation, Space, and Environmental Medicine*, 1104–1108.

Fry, R. J. M. (2002). Radiation in space: Risks estimates. *Radiation Protection Dosimetry*, 100, 475–477.

Gaca, P., Z. S. Zunic, J. W. Mietelski, E. Tomankiewicz, and M. P. R. Walingórski. (2005). Experimental results on the environmental samples collected around sites in South Serbia, Kosovo and Montenegro where DU weapons were deployed in 1999. In: *The Natural Radiation Environment VII. Proceedings of an International Symposium*, Rhodes, Greece, 20–24 May 2002 (Eds J. P. McLaughlin, S. E. Simopoulus, and F. Steinhäusler), Elsevier Radioactivity in the Environment Series Vol. 7 (M. S. Baxter, Series Editor), Amsterdam, pp. 1056–1063; ISSN 1569-4860, doi:10.1016/S1569-4860(04)07110-4.

Gamow, G. (1948a). The origin of elements and the separation of galaxies. *Physical Review*, 74, 505–506.

Gamow, G. (1948b). The evolution of the universe. *Nature*, 162, 680.

García-León, M., A. Martinez-Aguirre, R. Pertáñez, J. P. Bolivar, and R. García-Tenorio. (1995). Levels and behavior of natural radioactivity in the vicinity of phosphate fertilizer plants. *Journal of Radioanalytical and Nuclear Chemistry*, 197, 173–184. 10.1007/BFO2040229

George, A. C. (2008). World history of radon research and measurement from the early 1900s to today. In: *The Natural Radiation Environment VIII. Proceedings of an International Symposium*, Búzios, Rio de Janeiro, Brazil, 7–12 October 2007 (Eds A. S. Paschoa and F. Steinhäusler), American Institute of Physics Conference Proceedings 1034, New York, pp. 20–33.

George, A. C., and A. J. Breslin. (1980). The distribution of ambient radon and radon daughters in residential buildings in the New Jersey–New York areas. In: *Natural Radiation III* (Eds T. F. Gesell and W. M. Lowder), US Department of Energy Report CONF-780422 (DOE Symposium Series, Vol. 51, USDOE, Oak Ridge, TN).

Gesell, T. F. (1975). Occupational radiation exposure due to ^{222}Rn in natural gas and natural gas products. *Health Physics*, 29, 681–687.

Gesell, T. F. (1983). Background atmospheric ^{222}Rn concentration outdoors and indoors: A review. *Health Physics*, 45, 289–302.

Getseva, R. V. (1958). The nature of sedimentary metamorphic uranium mineralization. In: *The Geology of Uranium*, Supplement No. 6 of the Soviet Journal of Atomic Energy, Atomic Press, Moskow, 1957 (translated from Russian by Consultants Bureau, New York), Chapman and Hall, London, pp. 14–19.

Giammarchi, M. G., and L. Miramonti. (2006). Borexino: Geoneutrinos in Borexino. *Earth, Moon, and Planets*, 99, 207–220. 10.1007/s11038-006-9106-6.

Glasser, O. (1934). *Wilhem Conrad Röentgen and the Early History of the Röentgen Rays*. Thomas, Springfield, IL. See also, Dibner, B. (1968). *Wilhen Conrad Reöentgen and the Discovery of X-rays*. Watts, New York.

Glowiak, B. J., and J. M. Pacyna. (1980). Radiation dose due to atmospheric releases from coal/fired power stations. *International Journal of Environmental Studies*, 16, 23–28. 10.1080/00207238008709843.

Gockel, A. (1909). Radioactivity of the atmosphere. *Archives des Sciences Physiques et Naturelles*, 27, 248–255.

Gockel, A. W. F. E. (1911). LuftelektrischeBeobachtungen bei einer Ballonfahrt (Observations of atmospheric electricity during a balloon ride). *Physikalische Zeitschrift*, 12, 597.

Gockel, A. W. F. E. (1912). The Penetrating Radiation Present in the Atmosphere. *Jahrbuch der Radioaktivitat und Elektronik*, 9, 1–15.

Goldhagen, P., J. M. Clem, and J. W. Wilson. (2004a). Recent results from measurements of the energy spectrum of cosmic ray-induced neutrons aboard an ER-2 airplane and on the ground. *Advances in Space Research*, 32, 35–40.

Goldhagen, P., J. M. Clem, and J. W. Wilson. (2004b). The energy spectrum of cosmic-ray induced neutrons measured on an airplane over a wide range of altitude and latitude. *Radiation Protection Dosimetry*, 110, 387–392.

Goodman, C., K. C. Bell, and W. L. Whitehead. (1939). The radioactivity of sedimentary rocks and associated petroleum. *The American Geologist – Journal of the Mineralogical Society of America*, 24, 7.

Gosh, D., A. Deb, R. Sengupta, K. K. Patra, and S. Bera. (2007). Pronounced soil-radon anomaly – Precursor of recent earthquakes in India. *Radiation Measurements*, 42, 466–471. 10.1016/j.radmeas.2006.12.008.

Gosse, J. C., and F. M. Phillips. (2001). Terrestrial *in situ* cosmogenic nuclides: Theory and application. *Quaternary Science Review*, 20, 1475.

Goto, M., J. Moriizumi, H. Yamazawa, T. Iida, and W. Zhuo. (2008). *The Natural Radiation Environment VIII. Proceedings of an International Symposium*, Búzios, Rio de Janeiro, Brazil, 7–12 October 2007 (Eds A. S. Paschoa and F. Steinhäusler) American Institute of Physics Conference Proceedings 1034, New York, pp. 169–172.

Gunning, C., and A. G. Scott. (1982). Radon and thoron daughters in housing. *Health Physics*, 42, 427.

Guo, Q., J. Cheng, and Y. Chen. (2005). Dose evaluation of indoor thoron progeny in som areas in China. In: *The Natural Radiation Environment VII. Proceedings of an International Symposium*, Rhodes, Greece, 20–24 May 2002 (Eds J. P. McLaughlin, S. E. Simopoulus, and F. Steinhäusler), Elsevier, Amsterdam, pp. 506–511.

Hacker, B. C. (1987). *The Dragon's Tail: Radiation safety in the Manhattan Project, 1942–1946*. University of California Press, Berkeley, CA.

Hahn, O. (1958). The discovery of fission. *Scientific American*, CXCVIII, 76–78.

Hahn, O., and L. Meitner. (1935). *Naturwiss*, 23, 320.

Hahn, O., and E. Walling. (1938). The possibility of determining the geological age of Rubidium-containing minerals and rocks. In: *Geochronology: Radiometric Dating of Rockes and Minerals* (Ed. C. T. Harper), Douden, Hutchingson & Ross, Inc., Stroudsburg, PA (1973), pp. 220–223, (translated from Über die Möglichchkeit Geologischer Altersbestimmungen Rubidiumhalttiger mineralien undGesteine. Zeitschrift fur Anorganische und Allgemeine Chemie, 236, pp. 78–82.

Hajek, M., T. Berger, L. Summerer, W. Schöner, and N. Vana. (2005). Measurements and calculations of the radiation exposure of aircrew personnel on different flight routes. In: *The Natural Radiation Environment VII. Proceedings of an International Symposium*, Rhodes, Greece, 20–24 May 2002 (Eds J. P. McLaughlin, S. E. Simopoulus, and F.

Steinhäusler), Elsevier, Amsterdam, pp. 948–954 (OBS.: Measurements during solar maximum).

Hamlat, M. S., H. Kadi, S. Djeffal, and H. Brabini. (2003). Radon concentrations in Algerian oil and gas. *Applied Radiation and Isotopes*, 58, 125–130.

Haridasan, P. P., P. M. B. Pillai, R. M. Tripathi, and V. D. Puranaki. (2008). Thorium in ilmenite and its radiological implications in the production of titanium dioxide. *Radiation Protection Dosimetry*, 129, 381–385. 10.1093/rpd/ncm446.

Heaton, B., and J. Lambley. (1995). TENORM in the oil, gas and mineral mining industry. *Applied Radiation and Isotopes*, 46, 577–581.

Henrich, E., and A. Janssens. (2005). The European Union basic safety standards on natural radiation sources. In: *Proceedings of the 9th EAN ALARA Workshop* (Ed. M. Williams), 18–21 October, Augsburg, Germany.

Hess, V. F. (1912a). Über Beobachtungen der durchdringenden Strahlung bei sieben Freiballonfahrten. *Physikalische Zeitschrift*, 13, 1084.

Hess, V. F. (1912b). Messungen der durchdringenden Strahlung bei zwei Freiballonfahrten (Measurements of penetrating radiation during two free ballon flights). *Wiener Sitzungsberichte Berichte*, 120, 1575–1585.

Hess, V. F., and J. Eugster. (1949). *Cosmic Radiation and its Biological Effects*. Fordham University Press, Bronx, NY.

Hewlett, B. C., and O. E. Anderson, Jr. (1962). *History of the United States Atomic Energy Commission. The New World, 1939/1946*. The Pennsylvania State University Press, University Park, PA, 766pp.

Hooper, R. (2009). Italian authorities dismissed quake warning. *New Scientist*, April 6, 2:45 PM, http://www.newscientist.com/blogs/shortsharpscience/2009/04/italian-quake-was-predicted-an.html.

Horne, R. B., R. M. Thom, Y. Y. Shprits, N. P. Meredith, S. A. Glauert, A. J. Smith, S. G. Kanekai, D. N. Baker, M. J. Engebreston, J. L. Posch, M. Spasojevic, U. S. Inan, J. S. Pickett, and P. M. E. Decreau. (2005). Wave acceleration of electrons in the Van Allen radiation belts. *Nature*, 437, 227–230. 10.1038/nature03939.

Hosoda, M., M. Shimo, M. Sugiro, M. Furukawa, M. Fukushi, K. Minami, and K. Ejiri. (2008). Radon and thoron exhalation rate map of Japan. In: *The Natural Radiation Environment VIII. Proceedings of an International Symposium*, Búzios, Rio de Janeiro, Brazil, 7–12, October 2007 (Eds Anselmo S. Paschoa, and F. Steinhäusler), American Institute of Physics Conference Proceedings 1034, New York, pp. 177–180.

Hughe, J. C., and G. Brown. (1977). Two unusual minerals in a Nigerian Soil: (I) Fibrous kaolin; (II) Bastnaesite. *Cay Minerals*, 12, 319–329.

Hull, A. P. (1971). Radiation in perspective: Some comparisons of the environmental risks from nuclear and fossil-fuel power plants. *Nuclear Safety*, 12, 186–196.

Hull, A. P. (1974). Comparing effluent releases from nuclear and fossil-fuel power plants. *Nuclear Safety*, 17, 51–55.

Hull, C. D., and W. C. Burnett. (1996). Radiochemistry of Florida phosphogypsum. *Journal of Environmental Radioactivity*, 32, 213–238. 10.1016/0265-931X(95)00061-E.

Igarashi, G., S. Saeki, N. Takahata, K. Sumikawa, S. Tasaka, Y. Sasaki, M. Takahashi, and Y. Sano. (1995). Ground-water radon anomaly before the Kobe earthquake in Japan. *Science*, 269, 60–61. 10.1126/science.269.5220.60

Institute of Medicine (IOM) of the National Academies. (2008). *Epidemiologic Studies of Veterans Exposed to Depleted Uranium: Feasibility and Design Issues*. Committee on Gulf War and Health: Updated Literature Review of Depleted Uranium, The National Academy Press, 58pp. ISBN: 0-309-1207-1.

International Atomic Energy Agency (IAEA). (1975). *Convention on the Prevention of Marine Pollution by Dumping of Wastes and Other Matter*. The Definition Required by Annex I,

Paragraph 6, to the Convention, and Recommendation Required by Annex II, Section D. INFCIRC/205/Add 1, Vienna.

International Atomic Energy Agency (IAEA). (1978a). *Convention on the Prevention of Marine Pollution by Dumping of Wastes and Other Matter.* The Definition Required by Annex I, Paragraph 6, to the Convention, and Recommendation Required by Annex II, Section D. INFCIRC/205/Add 1/Rev 1, Vienna.

International Atomic Energy Agency (IAEA). (1978b). *The Oceanographic Basis of the IAEA Revised Definition and Recommendations Concerning High-Level Radioactivity Waste Unsuitable for Dumping at Sea.* TECDOC-210, Vienna.

International Atomic Energy Agency (IAEA). (1978c). *The Radiological Basis of the IAEA Revised Definition and Recommendations Concerning High-Level Radioactivity Waste Unsuitable for Dumping at Sea.* TECDOC-211, Vienna.

International Atomic Energy Agency (IAEA). (1978d). *Principles for Establishing Limits for the Release of Radioactive Materials into the Environment.* Safety Series No. 45, IAEA, Vienna, 91pp.

International Atomic Energy Agency (IAEA). (1985). *Principles for Limiting Releases of Radioactive Effluents into the Environment.* Safety Series No. 77, IAEA, Vienna, 32pp.

International Atomic Energy Agency (IAEA). (1988). *Principles for the Exemption of Radiation Sources and Practices from Regulatory Control.* IAEA Safety Series No. 89, IAEA, Vienna.

International Atomic Energy Agency (IAEA). (1995). *The Principles of Radioactive Waste Management.* IAEA Safety Series No. 111-F, STI/PUB/989. Vienna, 36pp.

International Atomic Energy Agency (IAEA). (1996a). *International Basic Safety Standards for Protection against Ionizing Radiation and for the Safety of Radiation Sources.* IAEA Safety Series No. 115, STI/PUB/996, Vienna, 48pp.

International Atomic Energy Agency (IAEA) (1996b). *Radiation Protection and the Safety of Radiation Sources.* IAEA Safety Series No. 120, STI/PUB/1000, Vienna, 36pp.

International Atomic Energy Agency (IAEA) (1999a). *Occupational Radiation Protection.* Safety Guide, IAEA Safety Standards Series, No. RS-G-1.1, STI/PUB/1081, Vienna, 83pp.

International Atomic Energy Agency (IAEA) (1999b). *Decommissioning of Medical, Industrial and Research Facilities.* Safety Guide, IAEA Safety Standards Series, No. WS-G-2.2, STI/PUB/1078, Vienna, 45pp.

International Atomic Energy Agency (IAEA) (1999c). *Assessment of Occupational Exposure due to Intake of Radionuclides.* Safety Guide, IAEA Safety Series, No. RS-G-1.2, STI/PUB/1077, Vienna, 95pp.

International Atomic Energy Agency (IAEA). (2000a). *Predisposal Management of Radioactive Waste, Including Decomissioning.* Safety Requirements, Safety Standards Series No. WS-R-2, STI/PUB/1089, 36pp.

International Atomic Energy Agency (IAEA). (2000b). *Management of Radioactive Waste from the Mining and Milling of Ores.* Safety Guide, IAEA Safety Standards Series No. WS-G-1.2, STI/PUB/1134, Vienna, 48pp.

International Atomic Energy Agency (IAEA). (2000c). *Regulatory Control of Radioactive Discharges to the Environment.* Safety Guide, IAEA Safety Standards Series No. WS-G-2.3, STI/PUB/1088, Vienna, 52pp.

International Atomic Energy Agency (IAEA). (2001). *Generic Models for Use in Assessing the Impact of Discharges of Radioactive Substance to the Environment.* Safety Report Series No. 19, IAEA, Vienna, 216pp.

International Atomic Energy Agency (IAEA). (2003a). *Extent of Environmental Contamination by Naturally Occurring Radioactive Material (NORM) and Technological Options for Mitigation.* IAEA Technical Report Series No. 419, STI/DOC/010/419, Vienna, 208pp.

International Atomic Energy Agency (IAEA). (2003b). *Radiation Protection and Management of Radioactive Waste in the Oil and Gas Industry*. IAEA Safety Report Series No. 34, STI/PUB/1171, Vienna, 139pp.

International Atomic Energy Agency (IAEA). (2003c). *Depleted Uranium*. IAEA, Werner Burkart Interview, http://www.iaea.or.worldatom/Press/Focus/DU.

International Atomic Energy Agency (IAEA). (2006). *Fundamental Safety Principles*. IAEA Safety Standards Series No. SF-1, STI/PUB/1273.

International Atomic Energy Agency (IAEA). (2007). *Radioecological Parameter Values –* Update of TRS 364 (IAEA/IUR) – Task Groups in the Termination Phase – Task Group Leaders: P. Calmon and G. Voigt.

International Commission on Radiation Protection (ICRP). (1983). *Radionuclide Transformations-Energy and Intensity of Emissions*. Report of Task Group of Committee 2 ICRP Publication 38, 1250pp.

International Commission on Radiation Protection, (ICRP). (1991). *Recommendations of the International Commission on Radiological Protection*. ICRP Publication 60.

International Commission on Radiological Protection (ICRP). (1993). *Protection against Radon-222 at Home and at Work*. ICRP Publication 65, *Annals of ICRP*, 23. Oxford, Pergamon Press.

International Commission on Radiological Protection (ICRP). (1997a). *Conversion Coefficients for use in Radiological Protection against External Radiation*. ICRP Publication 74, *Annals of the ICRP*, 26/3.

International Commission on Radiation Protection (ICRP). (1997b). *General Principles for the Radiation Protection of Workers*. ICRP Publication 75.

International Commission on Radiological Protection (ICRP). (1979). *Report of Committee 4 on Radionuclide Release into the Environment: Assessment of Doses to Man*. ICRP Publication 29, *Annals of the ICRP*, 2. Oxford Pergamon Press, 76pp.

International Commission on Radiological Protection (ICRP). (2004). *Relative Biological Effectiveness (RBE), Quality Factor (Q), and Radiation Weighting Factor (wR)*. ICRP Publication 92, *Annals of the ICRP*, 80pp.

International Commission on Radiation Units and Measurements (ICRU). (1994). *Gamma-Ray Spectrometry in the Environment*. ICRU Report 53.

Irvine, D., and D. J. C. Flower. (2005). Practical considerations for implementation of the EU Directive in the field of cosmic radiation exposure. In: *The Natural Radiation Environment VII. Proceedings of an International Symposium*, Rhodes, Greece, pp. 20–24 May 2002 (Eds J. P. McLaughlin, S. E. Simopoulus, and F. Steinhäusler), Elsevier Radioactivity in the Environment Series Vol. 7 (M. S. Baxter, Series Editor), Amsterdam, pp. 876–884, ISSN 1569–4860, doi:10.1016/S1569-480(04) 07110-4.

Ishikawa, T., S. Tokonami, Q. Sun, Y. Kobayashi, X. Min, and S. Yoshinaga. (2007). Preliminary results of indoos radon/thoron concentrations and terrestrial gamma doses in Gejiu, Yunnan, China. In: *The Natural Radiation Environment VIII. Proceedings of an International Symposium*, Búzios, Rio de Janeiro, Brazil, 7–12 October 2007 (Eds A. S. Paschoa and F. Steinhäusler), American Institute of Physics Conference Proceedings 1034, New York, pp. 173–176.

Jacobi, W. (1972). Activity and potential alpha-energy of ^{222}Radon and ^{220}Radon daughters in different air atmospheres. *Health Physics*, 22, 441–450.

Janssens, A. (1999). *Concepts of exemption and clearance in the EU Basic Safety Standards*. Joint NEA/IAEA/EC Workshop on Regulatory Aspects of Decommissioning, Rome, Italy, 19–21 May, pp. 148–163.

Jarvis, A. N., and L. Siu. (1981). *Environmental Radioactivity Laboratory Intercomparison Studies Program*. EPA-600/4-81-004, EPA, Las Vegas.

Jasinski, S. M. (2007). Phosphate rock. In: *2007 Minerals Yearbook, Phosphate Rock* (Advance Release), US Department of Interior, US Geological Survey (USGS), August 2008, 12pp.

Johnston, W. R. (2006). High-altitude nuclear explosions, http://www.johnston's Archive – Nuclear Weapons – files. Accessed on 11 February 2009.

Joly, J. (1908). On the radium content of deep sea-sea sediments. *Philosophical Magazine*, 16, 190.

Jonkers, G., F. A. Hartog, A. A. I. Knaepen, and P. F. J. Lancee. (1997). Characterization of NORM in the oil and gas production (E&P) industry. In: *Radiological Problems with Natural Radioactivity in the Non-Nuclear Industry, Proceedings of an International Symposium*, Amsterdam, The Netherlands, 8–10 September 1997, KEMA, Arhem.

Karam, P. A. and S. A. Leslie. (2005). Changes in the terrestrial natural radiation levels over the history of life. In: *The Natural Radiation Environment VII. Proceedings of an International Symposium*, Rhodes, Greece, 20–24 May 2002 (Eds J. P. McLaughlin, S. E. Simopoulus, and F. Steinhäusler), Elsevier, Amsterdam, pp. 107–117.

Karangelos, D. J., P. K. Rouni, N. P. Petropoulos, M. J. Anagnostakis, E. P. Hinis, and S. E. Simopoulos. (2005). Radioenvironmental survey of the Megalopolis power plants fly ash deposits. In: *Radioactivity in the Environment* (Eds J. P. McLaughlin, S. E. Simopoulos, F. Steinhäusler), Elsevier Radioactivity in the Environment (Ed. M. S. Baxter), Vol. 7, pp. 1025–1039; ISSN 1569-4860, doi:10.1016/S1569-4860(04)07110-4).

Karayigit, A. I., Y. Bulut, G. Karayigit, X. Querol, A. Alastueey, S. Vassilev, and C. Vassilev. (2006). Mass balance of major and trace elements in a coal-fired power plant. *Energy Sources, Part A: Recovery, Utilization, and Environmental Effects*, 28, 1311–1320. 10.1080/009083190910523.

Kenaway, M. A., and A. A. Morsy. (1991). Measurements of environmental radon-222 concentration indoor and outdoors in Egypt. *Nuclear Tracks and Radiation Measurements*, 19, 343–345.

Kerness, N., and Taber, A. (1997). *Neutron SEU Trends in Avionics*. NSREC97 Radiation Effects Data Workshop Record, IEEE 97TH8293.

Khan, H. A., M. Tufail, and A. A. Qureshi. (1990). Radon signals for earthquake prediction and geological prospection. *Journal of Islamic Academy of Sciences*, 3, 229–231.

Kharitonov, I. A., M. A. Rasko, S. V. Sepman, E. E. Terechtchenko, and A. M. Hedjelman. (2002). A source for measurement of the absolute intensities of ^{226}Ra gamma-radiation in equilibrium with decay products. *Applied Radiation and Isotopes*, 56, 37–40.

Kip, A. S. (1962). *Fundamental of Electricity and Magnetism*. New York, McGraw-Hill Book Company, 1st Ed., 352pp.

King, C.-Y. (1981). Do radon anomalies predict earthquakes? *Nature*, 293, 262.

Klausman, A. L. (1995). Effects of space flight on small portable computers. MSc Thesis, University of Houston-Clear Lake, Houston, TX.

Klerkx, J., B. Dehandschutter, A. Annunziatellis, A. Baccani, T. Bituh, I. Celikovic, G. Giotoli, M. Coltella, A. Demajo, S. Dogjani, V. Gavshin, N. Gradascevic, L. Hoxha, P. Jovanovic, L. Juasz, D. Kisic, S. Kolobova, J. Kovac, S. Lombardi, V. Matychenkov, M. Melgunov, S. Meng, A. Mihali, B. Petrinec, A. Poffijn, A. Popovic, D. Samek, A. Samsonova, L. Saracevic, P. Szerbin, P. Uhc, and Z. S. Zunic. (2007). Environmental impact of radioactivity in waste from the coal and aluminium industries in western Balkan countries. In: *Naturally Occurrin Radioactive Material (NORM V), Proceedings of an International Symposium*, Seville, Spain, 19–22 March, International Atomic Energy Agency, Vienna, pp. 467–478.

Kolb, W. A., and M. Wojcik. (1985). Enhanced radioactivity due to natural oil and gas production and related radiological problems. *The Science of the Total Environment*, 45, 77–84.

Komosa, A., S. Chiposwski, M. Klimek, and S. Chalupnik. (2004). Attempts on determination of radon exhalation rate from a waste dump of Bogdanka coal mine with use of the Picorad detectors. In: *Proceedings of International Conference on Naturally Occurring Radioactive Materials (NORM IV)*, Szczyrk, Poland, 17–21 May, IAEA-TECDOC-1472, Vienna, pp. 548–554.

Koperski, J., and N. Tsurikov. (1999). Impact of ARPANS-like legislation on mineral industry in Australia – the TENORM issue. In: *Proceedings of the Australian Radiation Protection Society Annual Conference*, Margaret River, Western Australia. Nick's TENORM Notes http://www.calytrix.biz/radlinks/tenorm/index/htm.

Korolkov, I. (2005). Yadovitaya Ataka [Poisonous Attack], Moskovskiye Novosti, No. 11, 18 March (DSTO).

Kozak, K., J. Swakoń, M. Paszkowski, R, Gradziński, J. Loskiewicz, M. Janik, J. Mazur, J. Bogacz, T. Horwacik, and P. Olko. (2005). Correlation between radon concentration and geological structure of the Kraków area. In: *The Natural Radiation Environment VII. Proceedings of an International Symposium*, Rhodes, Greece, 20–24 May 2002 (Eds J. P. McLaughlin, S. E. Simopoulus, and F. Steinhäusler) Elsevier, Amsterdam, pp. 464–469.

Krane, K. S. (1987). *Introductory Nuclear Physics*. John Wiley, New York, 3rd Edition (October 22, 1987), 864pp.

Kraner, H. W., G. L. Schroeder, and R. D. Evans. (1964). Measurements of the effects of atmospheric variables on radon-222 flux and soil-gas concentrations. In: *The Natural Radiation Environment. Proceedings of an International Symposium*, Houston, TX, April 10–13, 1963 (Eds John A. S. Adams, and Wayne M. Lowder), William Marsh Rice University by The University of Chicago Press, pp. 191–215.

Krauskopf, K. (1967). *Introduction to Geochemistry*. McGraw-Hill International Series in the Earth and Planetary Sciences, McGraw-Hill Book Company, New York, 721pp.

Kremers, H. E. (1949). The rare earth industry. *Transactions of the Electrochemical Society*, 96, 152–158.

Kvasnicka, J. (1996). Radiation protection in the offshore petroleum industry. *Proceedings of the 1996 International Congress on Radiation Protection (IRPA9)*, Vol. 4, 4-625–4-627.

Laborde, A. (1911). *Méthodes de measure employées en radioactivité*. Gathier-Villard, Paris.

Landa, E. R. (2004). Uranium mill tailings: Nuclear waste and natural laboratory for geochemical and radioecological investigations. *Journal of Environmental Radioactivity*, 77, 1–27. 10.1016/jenvrad.2004.01/030.

Lapp, R. E. (1949). Survey of nucleonics instrumentation industry. *Nucleonics*, 4, 101.

Lauria, D. C., and E. R. R. Rochedo. (2005). The legacy of monazite processing in Brazil. *Radiation Protection Dosimetry*, 114, 546–550.

Lave, L. B., and L. C. Freeburg. (1973). Health effects of electricity from coal, oil, and nuclear fuel. *Nuclear Safety*, 14, 409–428.

Learned, J. (2007). *Hano Hano: Hano Hano Summary*. Deep Ocean Anti-Neutrino Observatory Workshop (DOANOW), Hawai'i, http://www.phys.hawaii.edu/~sdye/doanow_talks/Learned_DOANOW_2007.ppt.

Leopold, K., and B. Michalik. (2007). Availability of radium isotopes and heavy metals from scales and tailings of Polish hard coal mining. In: *Naturally Occurring Radioactive Material (NORM V)*. Presented as Poster 3–7 at the NORM V International Symposium, Seville, Spain, 19–22. Available in CD-ROM format – not included in the NORM V Proceedings.

Lewis, R. K. (2006). A history of radon – 1470 to 1984. In: *16th National Conference on Radon*, Kansas City, MO, September 17–20, Conference of Radiation Control Program Directors, Inc., CRCPD Publication E-06-7, http://www.crcpr.org.

Lewis, B. J., G. I. Bennett, A. R. Green, A. Butler, M. Desormeaux, F. Kictching, M. J. McCall, B. Ellaschuk, and M. Pierre. (2005). Aircrew dosimetry using the predictive code for aircrew radiation exposure (PCAIRE). *Radiation Protection Dosimetry*, 116, 320–326.

Lewis, B. J., M. J. McCall, A. R. Green, L. G. I. Bennett, M. Pierre, U. J. Schrewe, K. O'Brien, and E. Felsberger. (2001). Aircrew exposure from cosmic radiation on commercial airline routes. *Radiation Protection Dosimetry*, 93(4), 293–314.

Libby, W. F. (1955). *Radiocarbon dating*. University of Chicago Press, Chicago. 2nd Ed.

Lin, W., and G. Harbottle. (1991). Gamma ray emission intensities of ^{226}Ra in equilibrium with its daughters. *Journal of Radioanalytical Nuclear Chemistry Letters*, 153, 137.

Lind, S. C. (1931). Origin of petroleum. *Science*, 73, 19.

Lind, S. C., and D. C. Bardwell. (1926). The chemical action of gaseous ions produced by alpha particles. IX. Saturated hydrocarbons. *Journal of the American Chemistry Society*, 48, 2335–2351.

Linsley, G. (1999). *International Guidance on the Removal of Regulatory Control from Materials Containing Radionuclides*. Joint NEA/IAEA/EC Workshop on Regulatory Aspects of Decommissioning, Rome, Italy, 19–21 May, pp. 140–147.

Lockart, Jr. (1964). Radioactivity of the Radon-222 and Radon-220 series in the air at ground level. In: *The Natural Radiation Environment. Proceedings of an International Symposium*, Houston, TX, April 10–13, 1963 (Eds John A. S. Adams and Wayne M. Lowder), William Marsh Rice University by The University of Chicago Press, pp. 331–344.

London Convention. (1993). The 1993 amendments. Adoption: 12 November 1993. Entry into force: 20 February 1994.

The amendments banned the dumping into sea of low-level radioactive wastes. In addition, the amendments:

- phased out the dumping of industrial wastes by December 1995
- banned the incineration at sea of industrial wastes.

Although all three disposal methods were previously permitted under the Convention, attitudes towards the use of the sea as a site of disposal of wastes have changed over the years.

In 1993 the Contracting Parties of the LC adopted a resolution calling for a moratorium on the sea dumping of low-level radioactive wastes. Later resolutions called for phasing-out of industrial waste dumping and on end to the incineration at sea of noxious liquid wastes

Lucas, Jr., H. F. (1964). A fast and accurate survey of technique for both Radon-222 and radium-226. In: *The Natural Radiation Environment. Proceedings of an International Symposium*, Houston, TX, April 10–13, 1963 (Eds John A. S. Adams, and Wayne M. Lowder), William Marsh Rice University and The University of Chicago Press, pp. 315–329.

Ludwig, G. H. (2004). The first explorers satellites. Invited talk at the Van Allen's 90th Birthday Celebration at the University of Iowa, 9 October. http://www.pw.physics.uiowa.edu/van90/ExplorerSatellites_LudwigOct2004.pdf.

Ludwig, T., I. Schäfer, and G. Seitz. (2008). Radiation exposure in the production and use of thoriated gas mantles. In: *Naturally Occurring Radioactive Material (NORM V), Proceedings of an International Symposium*, Seville, Spain, 19–22 March 2007, International Atomic Energy Agency, Vienna, pp. 71–79.

Luther, S. M., M. J. Dudas, and P. M. Rutherford. (1993). Radioactivity and chemical characteristics of Alberta phosphogypsum. *Water, Air and Soil Pollution*, 69, 277–290.

Machado, N., and G. Gauthier. (1996). Determination of $^{207}Pb/^{206}Pb$ ages on zircon and monazite by laser-ablation ICPMS and application to a study of sedimentary provenance and metamosphism in southeastern Brazil. *Geochimica et Cosmochimica Acta*, 60, 5063–5073. 10.1016/S0016-7037(96)00287-6.

Mahesh, H. M., D. N. Avadhani, and K. Siddappa. (2005). Concentration, distribution and transportation of ^{222}Rn and its decay products in the environment of coastal Karnataka and Kaiga in southwest India. In: *The Natural Radiation Environment VII. Proceedings of an International Symposium*, Rhodes, Greece, 20–24 May 2002 (Eds J. P. McLaughlin, S. E. Simopoulus, and F. Steinhäusler), Elsevier, Amsterdam, pp. 542–553.

Manyama Makweba, M., and E. Holm. (1993). The natural radioactivity of rock phosphates, phosphatic products and their environmental implications. *Science of the Total Environment*, 133, 99–110.

Markovskii, N. I. (1978). *Paleogeographic Principles of Oil and Gas Prospecting* (translated from Russian by R. Tateruk-Schneider (Ed. R. Amoils)). Wiley, New York, 256pp.

Martignano, M., and R. Harboe-Sorensen. (1995). IBM ThinkPad radiation testing and recovery during EUROMIR missions. *IEEE Transactions on Nuclear Science*, NS-42 (6), 2004–2009.

Martin, J. E., E. D. Harward, and D. T. Oakley. (1970). Comparison of radioactivity from fossil-fuel and nuclear power plants. In: Environmental Effects of Producing Electric Power, Part I. Appendix 14 – Joint Committee on Atomic Energy, 91st Congress of the US, 1st Session, Washington, D.C., November 1969, pp. 773–809. Published also as – Martin J. E. et al., Radioactivity from fossil-fuel and nuclear plants. *International Atomic Energy Agency (IAEA) Symposium SM-146/19*, New York, NY, 10–14 August.

Masarik, J., and J. Beer. (1999). Simulation of particle fluxes and cosmogenic nuclide production in the Earth's atmosphere. *Journal of Geophysical Research.*, 104, 12099–12111.

McBride, J. P., R. E. Moore, J. P. Whitherspoon, and R. E. Blanco. (1977). *Radiological Impact of Airborne Effluents of Coal-Fired and Nuclear Power Plants.* Oak Ridge National Laboratory (ORNL), Health and Safety Research Division, Energy research and Development Administration (ERDA), Contract No. W-7405-eng-26, ORNL-5315, 51 pp.

McDonald, A. (2008). International Law and depleted uranium weapons: A precautionary approach. *Report of the Expert Meeting on the Risks of Depleted Uranium Use in Weapon Systems*, Amersfoort, The Netherlands, February 14. Laka Foundation, NVMP.

McLaughlin, J. P. (1988). *Radon in Indoor Air.* Commission of the European Communities (Joint Research Centre, Ispra, Italy). Report EUR11917EN Luxemburg, November 1988.

McLaughlin, J. P. (2005). The exposure potential of depleted uranium in the environment – a review. In: *The Natural Radiation Environment VII. Proceedings of an International Symposium*, Rhodes, Greece, 20–24 May 2002 (Eds J. P. McLaughlin, S. E. Simopoulus, and F. Steinhäusler), Elsevier, Amsterdam, pp. 95–104.

McLennan, J. C. (1904). *On the Radioactivity of Mineral Oils and Natural Gas.* Paper presented at International Electric Congress, St. Louis, MO.

McLennan, J. C., and E. F. Burton. (1902). *Physical Review (Series I)*, 16, 173–192 (1903) (Issue 3 – March 1903).

McNulty, G. S. (2008). Production of titanium oxide. In: *Naturally Occurring Radioactive Material (NORM V), Proceedings of an International Symposium*, Seville, Spain, 19–22 March 2007, International Atomic Energy Agency, Vienna, pp. 169–188.

Miller, E. W. (1990). Radon hazard: A historical perspective. In: *Environmental Radon: Occurrence, Control and Health Hazards* (Eds S. K. Manjumdar, R. F. Schmalz, and E. Willard Miller). The Pennsylvania Academy Press, Easton, PA, pp. 1–12.

Millikan, R. A. (1911). The isolation of an ion, a precision measurement of its charge, and the correction of Stokes's law. *Phys. Rev. (Series I)*, 32, 349–397.

Millikan, R. A. (1913). Concerning the probable value of the friction co-efficient of air. *Annalen Der Physik*, 41, 759–766.
Millikan, R. A. (1993). *The Electron.* University of Chicago Press, Chicago.
Minato, S. (2006). Distribution of terrestrial γ ray dose rates in Japan. *Journal of Geography*, 115, 87–95.
Montagne, C., J. P. Donne, D. Pelcot, V. D. Nguyen, P. Bouisset, and G. Kerlau. (1993). In-flight radiation measurements aboard French airliners. *Radiation Protection Dosimetry*, 48, 79–83.
Mora, J. C., J. A. Corbacho, B. Robles, A. Baeza, D. Cancio, and A. M. Suarez. (2008). Methodology used in radiological assessment of a coal-fired power plant. In: *The Natural Radiation Environment VIII. Proceedings of an International Symposium*, Búzios, Rio de Janeiro, Brazil, 7–12 October 2007 (Eds A. S. Paschoa and F. Steinhäusler), American Institute of Physics Conference Proceedings 1034, New York, pp. 311–314.
Moravanska, H., and A. Laciok. (2004). Presence of technologically-enhanced naturally-occurring radioactive materials in the Czech Republic. In: *Proceedings of International Conference on Naturally Occurring Radioactive Materials (NORM IV)*, Szczyrk, Poland, 17–21 May, IAEA-TECDOC-1472, Vienna, pp. 202–212.
Morel, J., M. Etcheverry, and J. L. Picolo. (1998). Emission probabilities of the main γ-rays following the decay of ^{222}Rn and daughters. *Applied Radiation and Isotopes*, 49, 1387.
Morkunasi, G., L. Pilkytei, and D. Ereminas. (2003). Evaluation of exposure to cosmic radiation of flight crews of Lithuanian airlines. *International Journal of Occupational Medicine and Environmental Health*, 16, 161–167.
Mugrauer, O., D. E. Becker, P. Guglhoer. (1993). Strahlenexposition durch den Umgang mit Reststoffen aus der Kohleverbrennung und den daraus hergestellten Verbrauchsguetern. *Proceedings VGB-Conference Forschung in der Kraftwerkstechnik*, VGB-TB 231, VGB Krfatwerkstechnik GmbH, Essen.
Mutscheller, A. (1925). Physical standards of protection against Roentgen – Ray dangers. *American Journal of Radiology*, 13, 65–70.
National Academy of Sciences – National Research Council (NAS-NRC). (1989). *The Nuclear Weapons Complex: Management for Health, Safety, and the Environment.* National Academy Press, Washington, D.C., 157pp.
National Academy of Sciences – National Research Council (NAS-NRC). (1999). *Evaluation of Guidelines for Exposures to Technologically Enhanced Naturally Occurring Radioactive Materials.* National Academy Press, Washington, D.C., 281pp.
National Council on Radiation Protection (NCRP). (1984). *Radiological Assessment: Predicting the Transport, Bioaccumulation and uptake by Man of Radionuclides Released to the Environment.* NCRP Report No. 76, Bethesda, MD.
National Council on Radiation Protection and Measurements (NCRP). (1987). *Exposure of the Population in the United States and Canada from Natural Background Radiation.* NCRP Report No. 94, Bethesda, MD.
National Council on Radiation Protection and Measurements (NCRP). (1993). *Radiation Protection in the Mineral Extraction Industry.* NCRP Report No. 118, Bethesda, MD.
National Council on Radiation Protection and Measurements (NCRP). (2002). *Managing Potentially Radioactive Scrap Metal.* NCRP Report No. 141, Bethesda, MD.
National Research Council (NRC) of the National Academies. (2006). *Space Radiation Hazards and the Visions of Space Exploration*, Report of a Workshop, National Academy Press, Washington, D.C., 105pp.
Nazaroff, W. W., B. A. Moed, and R. G. Sextro. (1988). Soil as a source of indoor radon generation, migration and entry. In: *Radon and Its Decay Products in Indoor Air, An*

International Symposium (Eds W. A. Nazaroff and A. V. Nero, Jr.). Wiley, New York, pp. 57–112.

Negarestani, A., S. Setayeshi, M. Ghannadi-Maragheh, and B. Arashe. (2001). Layered neural networks based analysis of radon concentration and environmental parameters in earthquake prediction. *Journal of Environmental Radioactivity*, 62, 225–233. 1016/S0265-931X(01)00165-5.

Negarestani, A., S. Setayeshi, M. Ghannadi-Maragheh, and B. Arashe. (2003). Estimation of the radon concentration in soil related to the environmental parameters by a modified Adaline neural network. *Applied Radiation and Isotopes*, 58, 269–273.

Nelson, W. R. and T. M. Jenkinns. (Eds) (1980). *Computer Techniques in Radiation Transport and Dosimetry*. Plenum Press, New York, 521pp.

Nero, A. V., Jr.. (1988). Radon and its decay products in indoor air: An overview. In: *Radon and Its Decay Products in Indoor Air, An International Symposium* (Eds W. A. Nazaroff and A. V. Nero, Jr.). Wiley, New York, pp. 1–56.

Nero, A. V., Jr., M. B. Schehr, W. W. Nazarof, and K. L. Revzan. (1986). Distribution of airborne radon-222 concentrations in US houses. *Science*, 234, 992–997.

Neutrino Geoscience. (2008). A Workshop on Neutrino Geoscience, September 17–19, 2009, at Sudbury Neutrino Observatory Laboratory (SNOLab) Sudbury, ON, Canada.

Nicanov, V. F. (1969). *Geochemistry*, 199pp. (translated from *Doklady Akademii Nauk SSSR*, 188, 1148).

O'Brien, K. (1973). *LUIN, A Code for the Calculation of Cosmic Propagation in the Atmosphere*. Report HASL-275, Health and Safety Laboratory.

O'Brien, K. (1978). *LUIN, A Code for the Calculation of Cosmic Ray Propagation in the Atmosphere* (update of HASL-275), EML-338.

O'Brien, K. (2005). The theory of cosmic-ray and high-energy solar-particle transport in the atmosphere. In: *The Natural Radiation Environment VII. Proceedings of an International Symposium*, Rhodes, Greece, 20–24 May 2002 (Eds J. P. McLaughlin, S. E. Simopoulus, and F. Steinhäusler), Elsevier, Amsterdam, pp. 29–44.

O'Brien, K. (2008). Cosmic-rays in the solar system. In: *The Natural Radiation Environment VIII. Proceedings of an International Symposium*, Búzios, Rio de Janeiro, Brazil, 7–12 October 2007 (Eds A. S. Paschoa and F. Steinhäusler), American Institute of Physics Conference Proceedings 1034, New York, pp. 86–92.

O'Brien, K., W. Friedberg, H. H. Sauer, and D. F. Smart. (1996). Atmospheric cosmic rays and solar energetic particles at aircraft altitudes. In: *The Natural Radiation Environment VI, Proceeding of an International Symposium* (Ed. P. K. Hopke), Montreal, Que., Canada, 5–9 June, 1995. Published in *Environment International*, 22 (Suppl. 1), S9–S44.

O'Brien, K., and S. Skalski. (1996). Victor Hess: Taking natural radiation measurements to new heights. In: *The Natural Radiation Environment VI, Proceedings of an International Symposium* (Ed. P. K. Hopke), Montreal, Que., Canada, 5–9 June, 1995. Published in *Environment International*, 22(Suppl. 1), S121–S123.

O'Brien, K., D. F. Smart, M. A. Shea, E. Felsberger, U. Schrewe, W. Friedberg, and K. Copeland. (2003). World-wide radiation dosage calculations for air crew members. *Advances in Space Research*, 31, 835–840.

Oka, M., M. Shimo, S. Tokonami, A. Sorimachi, H. Takahashi, and T. Ishikawa. (2008). In: *The Natural Radiation Environment VIII. Proceedings of an International Symposium*, Búzios, Rio de Janeiro, Brazil, 7–12 October 2007 (Eds A. S. Paschoa and F. Steinhäusler), American Institute of Physics Conference Proceedings 1034, New York, pp. 189–192.

Oliveros, V., L. Aguirre, D. Morata, A. Somonetti, M. Vergara, M. Belmar, and S. Calderón. (2008). Geochronology of very low-grade Mesozoic Andean metabasite; an approach through the K-Ar, 40K-39Ar and U-Pb LA-MC-ICP-MS methods. *Journal of the Geophysical Society*, 165(Part 2), 579–584.

Olsen, J., P. E. Becher, P. B. Fynbo, P. Raaby, and J. Schultz. (1993). Neutron-induced single event upsets in static RAMs observed at 10 km flight altitude. *IEEE Transactions on Nuclear Science*, NS-40(2), 74–77.

Olson, G. (1983). Gamma ray intensities for ^{226}Ra and daughters, and ^{235}U. *Nuclear Instruments and Methods*, 206, 313.

Olson, J. C., D. R. Shawe, L. C. Pray, W. N. Sharpe, and D. F. Hewlett. (1954). Geology of the Rare Earth Deposits of the Mountain Pass district, San Bernardino County, CA, US Geological Survey (USGS), 261.

Omar, M., H. M. Ali, M. P. Abu, K. M. Kontol, Z. Ahmad, S. H. S. S. Ahmad, I. Sulaiman, and R. Hamzah. (2004). Distribution of radium in oil and gas industry wastes from Malaysia. *Applied Radiation and Isotopes*, 60, 779–782. 10.1016/j.apradiso.2004.01.005.

OSPAR. (1995). *Activities of the Oslo and Paris Commissions September 1992–June 1995, Implementation of the 1992, 1993, and 1994 OSPAR Action Plans.* Annual Reports. OSPAR, Paris; 1995. Report No. 1995.

O'Sullivan, D, L. Tommasino, H. Schraube, R. Grillmaier, D. T. Bartlett, L. Lindborg, W. Heinrich, and M. Silari. (1999). *Investigations of Radiation Fields at Aircraft Altitudes.* Final Report of European Commission Contract No. F14P-CT950011 (Dublin Institute for Advanced Studies – DIAS: Dublin).

Padamanabhan, V. T. (1987). All within limits. Radioactive waste disposal at IRE. *Economical and Political Weekly*, 22, 379–383.

Paridaens, J., and H. Vanmarcke. (2005). Radiological impact of flue gas purification in a coal fired power plant in Belgium, In: *Radioactivity in the Environment* (Eds J. P. McLaughlin, S. E. Simopoulos, and F. Steinhäusler), Elsevier (M.S. Baxter, Series Editor); Vol. 7, pp. 967–972, ISSN 1569-4860/doi:10.1016/S1569-4860(04)07110-4).

Parizot, E. (2007). *Cosmic-Rays: An Overview.* Lecture presented at the International School on Astro-Particle Physics (ISAPP School), Seillac, France, 4–15 June.

Parker, H. M., and W. C. Roesch. (1963). Units, radiation: Historical development. In: *The Encyclopedia of X-Rays and Gamma Rays* (Ed. G. L. Clark). Reihold, New York.

Paschoa, A. S. (1981). Old and new versions of the alpha radiation theory of petroleum origin. In: *Natural Radiation Environment. Proceedings of an International Symposium*, Bombay, India, January19–23 (Eds K. G. Vohra, K. C Pillai, U. C. Mishra, S. Sadasivan), Wiley Eastern Limited, New Delhi, pp. 651–657

Paschoa, A. S. (1988). Uncertainty analysis for ranking parameters in environmental mathematical modeling. In: *Methods for Assessing the Reliability of Radioactive Transfer Model Predictions, Proceedings of an International Workshop* (Ed. G. Desmet), Athens, Greece, 5–9 September, 1987, Commission of the European Communities, Elsevier Science Publishers, New York, NY, pp. 331–337.

Paschoa, A. S. (1990). Nuclear and atomic analytical techniques in environmental studies in South America. In: *Nuclear Analytical Methods in the Life Sciences* (Eds R. Zeisler and V. P. Guinn). Humana Press, Clinton, NJ. See also *Biological Trace Element Research*, Vol. 26., pp. 647–660.

Paschoa, A. S. (1993). Overview of environmental and waste management aspects of the monazite cycle. *Radiation Protection in Australia*, 11, 170–173.

Paschoa, A. S. (1994). The monazite cycle in Brazil: Past, present, and future. In: *Actinide Processing: Methods and Materials. Proceedings of an International Symposium*, San Francisco, CA, 28 February–3 March, 1994 (Eds B. Mishra and W. A. Averill), A publication of TMS – Minerals – Metals – Materials, pp. 323–338.

Paschoa, A. S. (1997). Radioactivity in the twentieth century and some of its implications for the third millenium. In: *Globalization of Nuclear Activities, an International Symposium*, June 15–18, 1997, Rio de Janeiro, Brazil. Proceedings available as CD-ROM by Latin American Section of the American Nuclear Society.

Paschoa, A. S. (2000a). More than forty years of studies of natural radioactivity in Brazil. *Technology*, 7, 192–212.

Paschoa, A. S. (2000b). The origins of the metrology of ionizing radiation. *Proceedings of Metrologia 2000*, São Paulo, Brazil, 4–7 December 2000. An international event organized by the Brazilian Society of Metrology.

Paschoa, A. S. (2001). Low gamma counting for measuring NORM/TENORM with a radon reducing system. *Radiation Physics and Chemistry*, 61, 501–504.

Paschoa, A. S. (2002). TENORM wastes and the potential alpha radiation dose to aquatic biota. In: *Technologically Enhanced Natural Radiation (TENR II). Proceedings of an International Symposium*, Rio de Janeiro, Brazil, 12–17 September 1999. IAEA-TECDOC-1271, IAEA, Vienna, pp. 92–96.

Paschoa, A. S. (2003). Radiological impact due to oil and gas extraction. In: *World Markets Research Centre (WMRC) Business Briefing*, Exploration and Production Report, 2–5, January 2003.

Paschoa, A. S. (2004). Gamma spectrometry – A simple probe for measuring soils with DU, NU and LEU. *Radiation Physics and Chemistry*, 71, 803–804.

Paschoa, A. S. (2008). Rapporteur summary on extraction and processing of rare earths. In: *Naturally Occurring Radioactive Material (NORM V). Proceedings of an International Symposium*, Seville, Spain, 19–22 March 2007, International Atomic Energy Agency, Vienna, pp. 3197–3323.

Paschoa, A. S. (2009). NORM from the monazite cycle and from the oil and gas industry: Problems and tentative solutions. In: *Radioecology and Environmental Radioactivity. Proceedings of an International Conference*, Bergen, Norway, 15–20 June, 2008. *Radioprotection*, 44, 957–962.

Paschoa, A. S., and G. B. Baptista. (1978). Method for calculation of upper limit internal alpha dose rates to aquatic organisms with application to plutonium-239 in plankton. *Heath Physics*, 35, 404–405.

Paschoa, A. S., G. B. Baptista, M. E. Wrenn, and M. Eisenbud. (1981). Dosimetry of natural and man-made alpha emitters in plankton. In: *Impacts of Radionuclide Releases into the Marine Environment. Proceedings of an International Symposium*, jointly organized by the International Atomic Energy Agency and the OECD Nuclear Energy Agency, Vienna, 6–10 October 1980, IAEA-SM-248/140, Vienna, pp. 695–716.

Paschoa, A. S., and B. M. Dantas. (2008). *An Attempt to Evaluate the Risks Associated with Radiological Terror.* Presented at the International Conference, 5th Dresden Symposium – Hazards – Detection and Management, March 03–07.

Paschoa, A. S., and K. Dias da Cunha. (2008). A critical look at NORM in the monazite cycle. In: *The Natural Radiation Environment VIII. Proceedings of an International Symposium*, Búzios, Rio de Janeiro, Brazil, 7–12 October 2007 (Eds A. S. Paschoa and F. Steinhäusler), American Institute of Physics Conference Proceedings 1034, New York, pp. 119–123.

Paschoa, A. S., M. Eisenbud, and J. M. Godoy. (1993). Radium availability in monazite sands from two distinct sites. *Radiation Protection in Australia*, 11, 162–164.

Paschoa, A. S., and J. M. Godoy. (2002). The areas of high natural radioactivity and TENORM wastes. In: *High Levels of Natural Radiation and Radon Areas: Radiation Dose and Health Effects. Proceedings of an International Conference*, Munich, Germany, September 4–7, 2000, pp. 3–8.

Paschoa, A. S., F. L. C. Moutella, M. T. Facchinetti, L. A. T. Rovere, O. Y. Mafra, and H. Evangelista. (2003). Calibrating a low gamma counting system for measuring depleted and low enriched uranium. In: *XVII IMEKO World Congress, Metrology in the 3rd Millenium, Proceedings of an International Congress*, June 22–27, 2003, Dubrovnik, Croacia, pp. 191–194.

Paschoa, A. S., and A. W. Nóbrega. (1981). Non-nuclear mining with radiological implications. In: *Radiation Hazards in Mining: Control, Measurements, and Medical Aspects.Proceedings of an International Conference*, Golden, CO, October 4–9, 1981 (Ed. M. Gomez), Society of Mining Engineers of American Institute of Mining, Metallurgical, and Petroleum Engineers, Inc., New York, NY, pp. 82–88.

Paschoa, A. S., and J. Pohl-Rülling. (2005). Early thoron concentration levels in Guarapari: Dosimetric implications. In: *The Natural Radiation Environment VII. Proceedings of an International Symposium*, Rhodes, Greece, 20–24 May 2002 (Eds J. P. McLaughlin, S. E. Simopoulus, and F. Steinhäusler), Elsevier Radioactivity in the Environment Series Vol. 7 (M. S. Baxter, Series Editor), Amsterdam, pp. 678–686, ISSN 1569-4860, doi:10.1016/S1569-480(04)07110-4.

Paschoa, A. S., J. A. Torrey, and M. E. Wrenn. (1985). Reducing radon exhalation from covered tailings: Optimization or cost-effectiveness?. *The Science of the Total Environment*, 45, 187–194.

Paschoa, A. S., M. E. Wrenn, and M. Eisenbud. (1979). Natural radiation dose to *Gammarus* from Hudson River. *Radioprotection*, 14, 99–115.

Paschoa, A. S., M. E. Wrenn, and J. A. Torrey. (1984). A mathematical model of indoor radon and daughters. *Radiation Protection Dosimetry*, 7, 139–142.

Pauli, W. (1930) – "Dear radioactive Ladies and Gentlemen! As the bearer of these lines, for whom I ask your gracious attention, will explain to you in more detail, I have faced with the "false" statistics of the N-14 and Li-6 nuclei as well as the continuous β-spectrum, stumbled upon a desperate remedy. Namely, the possibility that in the nucleus there could exist electrically neutral particles, which I will call neutrons, which have a spin of one-half and obey the exclusion principle and in addition also differ from light quanta in that they do not travel with the speed of light. The mass of the neutron must be of the same order of magnitude as the mass of the electron and in any case not larger than 0.01 proton mass. The continuous beta spectrum would then be understandable assuming that in β-decay a neutron is emitted along with the electron in such a way that the sum of the energies of neutron and electron is constant. ...

***At the moment I don't trust myself enough to publish anything about this idea, and turn confidently to you, dear radioactives, with the question of how one might experimentally prove such a neutron it its penetrating ability is similar to about 10 times that of γ-radiation. I admit that my remedy may at first seem only slightly probable, because if neutrons do exist they should certainly have been observed long ago! But, nothing ventured, nothing gained, and the gravity of the situation with the continuous beta spectrum is illustrated by a statement of my respected predecessor in this office. Herr [Peter] Debye, who told me recently in Brussels: "Oh, it is best not to think about it at all. Like the new taxes!" Thus one should seriously discuss every means of salvation. Therefore, dear radioactives, test and decide! Unfortunately, I cannot appear in Tübingen in person, since I am indispensible here due to a ball which will take place the night of December 6 to 7 in Zurich. With many greetings to you all, your more humble and obedient servant, W. Pauli." – This is a reproduction of the translation of the letter as it appears in Sime, R. L. (1996). *Lise Meitner: A Life in Physics* (California Studies in the History of Science, 11), A Centennial Book. University of California Press, Berkeley 94720, 540 pp/www.ucpress.eduwww.ucpress.edu, 107–108.

Pellicioni, M. (1998). Fluence to dose equivalent conversion data and radiation weighting factors for high energy radiation. *Radiation Protection Dosimetry*, 77, 159–170.

Penfold, J. S. S., K. R. Smith, M. P. Harvey, and S. F. Mobbs. (1998). Assessment of the radiological impact of coal-fired power stations in the United Kingdom. *Proceedings of the 2nd International Symposium (NORM II)*, Krefeld (Germany), November 10–13, pp. 67–71.

Penzias, A. A., and R. W. Wilson. (1965). A measurement of excess antenna temperature at 4080Mc/s. *Astrophysical Journal*, 142, 419–421.

Perel'man, A. I. (1977). Geochemistry of elements in the supergene zone. (Translated by Rita Teteruk-Schneider from the Russian 1972 edition) (Ed. A. A. Levinson), Keter Publishing House Ltd., 266pp.

Pettersson, H. (1930). Teneur en radium des depóts de mer profonde [Princesse-Alice et Challenger]. Résultats de Compagnes Scientifiques Albert 1er Prince de Monaco, Fasc. LXXXI, 1–50.

Pierce, A. P., J. W. Mytton, and G. B. Gott. (1955). Radioactive elements and their radioactive daughter products in the Texas Panhandale and oil and gas fields in the United States. In: *Contributions to the Geology of Uranium and Thorium by the United States Geological Survey and Atomic Energy Commission, an International Conference on the Peaceful Uses of Atomic Energy*, Geneva, Switzerland, pp. 527–532.

Piggot, C. S. (1933). Radium content of ocean-bottom sediments. *American Journal of Sciences*, 25, 229–238.

Pillai. P. M. B. (2008). Naturally occurring radioactive material (NORM) in the extraction and processing of rare earths. In: *Naturally Occurring Radioactive Material (NORM V). Proceedings of an International Symposium*, Seville, Spain, 19–22 March 2007, International Atomic Energy Agency, Vienna, pp. 197–221.

Planinić, J., V. Radolić, and V. Vuković. (2004). Radon as an earthquake precursor. *Nuclear Instruments and Methods in Physics Research A*, 530, 568–574. 10.1016/j.nima.2004.04.209.

Plastino, W., F. Bella, P. G. Catalano, and R. Di Giovambattista. (2002). Radon groundwater anomalies related to Umbra-Marche, September 26, 1997, earthquakes. *Geofisica International*, 41, 369–375.

Pontecorvo, B. (1942). Radioactivity analyses of oil well samples. *Geophysics*, 7, 90. doi:10.1190/1.1444997. Q&A: Polonium-210, 2006. *Chemistry World*. Royal Society of Chemistry (27 November 2006). http://www.rsc.org/chemistryworld/News/2006/November/27110601.asp. Last accessed on 20 February 2009.

Povinec, P. P., M. S. Baxter, J. Gayol, L. Liong Wee Kwong, I. Osvath, E. M. Scott. (1996a). *Global Marine Radioactivity Database (GLOMARD) – Radioactivity of the Artic Seas*. Report R4/96, IAEA-MEL, Monaco.

Povinec, P. P., A. Äarkrog, K. O. Buesseler, R. Delfanti, K. Hirose, G. H. Hong, T. Ito, H. D. Livingston, H. Nies, V. E. Noshkin, S. Shima, and O. Togawa. (2005). ^{90}Sr, ^{137}Cs, ^{239,240}Pu concentration surface water time series in Pacific and Indian Oceans – WOMARS results. *Journal of Environmental Radioactivity*, 81, 63–87.

Povinec, P. P., S. Fowler, and M. S. Baxter. (1996b). Chernobyl & the marine environment: The radiological impact in context. *International Atomic Energy Agency Bulletin*, 1, 18–22.

Preston, A. (1975). The radiological consequences of releases from nuclear facilities to the aquatic environment. In: *Impacts of Nuclear Releases into the Aquatic Environment. Proceedings of an International Symposium*, Otaniemi, Finland, 30 June–4 July 1975, IAEA, Vienna, pp. 3–23.

Privault, M. (1934). Les rayon x au laboratoire, a l'hopital, a l'usine. Librairie J.B. Bailiére et Fils, Paris, 204pp.

Puch, K. -H., R. Bialucha, and G. Keller. (2005). Naturally occurring radioactivity in industrial by-products from coal-fired power plants, from municipal waste incineration and from the iron- and steel-industry. In: *Radioactivity in the Environment* (Eds J. P. McLaughlin, S. E. Simopoulos, and F. Steinhäusler), Elsevier (M. S. Baxter, Series Editor), Vol. 7, pp. 996–1008, ISSN 1569-4860, doi:10.1016/S1569-4860(04)07110-4).

Quade, U., S. Thierfeldt, and S. Wörlen. (2003). Release of residues from melting NORM-contaminated steel scrap – A German approach. In: *Waste Management 2003 Conference (WM'03), Proceedings of an International Conference*, Tucson, AZ, February 23–27.

Raabe, O. G. (2002). According with information given by Otto G. Raabe on 25 November 2002 to the site http://www.vanderbilt.edu/radsafe: The term depleted uranium is a scientific term rather than a political statement. Originally, in the Manhattan Project in the 1940s it was called "depletalloy" to avoid the term uranium. The term "depleted" means that the fissile U-235 isotope has been reduced. The terminology is accurate, descriptive, and appropriate.

Raghavan, R. S. (2007). Hyper Scintillation Detector –HSD, Homestake: Large (50–200 kT) Liquid Scintillation Detector For keV–GeV Multidisciplinary Science-Hyper Scinitillation Detector – HSD, NSF-DUSEL Workshop, Washington D.C., November 3; http://neutrino.lbl.gov/~dusel/HomestakeWorkshopApril2008/HSD-Raghavan-1.ppt.

Ramola, A. R. C., M. Singh, A. S. Sandhu, S. Singh, and H. S. Virk. (1990). Use of radon gas as earthquake precursor. *Nuclear Geophysics*, 4, 275–287.

Ranft, J. (1972). Estimation of radiation problems around high energy accelerators using calculations of the hadronic cascade in matter. *Particle Accelerators*, 3, 129.

Rasonlonjatovo, D. A. H., H. Suzuki, N. Hirabayashi, T. Nunomiya, T. Nakamura, and N. Nakao. (2002). Measurements for the dose-rates of the cosmic-ray components on the ground. *Journal of Radiation Research*, 43(Suppl.), S27–S33.

Regener, E. (1909). On counting the alpha particles by scintillation and on the size of the electrical elementary quanum (in German), Sitzungsberichte der Konglich Preussischen Akademie Der Wissenschaften, 948–965.

Reines, F., and C. L. Cowan, Jr. (1956). The neutrino. *Nature*, 178, 446.

Reitz, G. (1993). Radiation Environment in the Stratosphere. *Health Physics*, 48, 5.

Revelle, R., and H. E. Suess. (1957). Carbon dioxide exchange between atmosphere and ocean and the question of an increase of atmospheric CO_2 during the past decades. *Tellus*, 9, 18–27.

Rodríguez Martínez, A. (2008). Radiological control of metal scrap. In: *Naturally Occurring Radioactive Material (NORM V). Proceedings of an International Symposium*, Seville, Spain, 19–22 March 2007, International Atomic Energy Agency, Vienna, pp. 327–337.

Rogers, V. C., and K. K. Nielsen. (1991). Multiphase radon generation and transport in porous media. *Health Physics*, 60, 261–265.

Roser, F. X., G. Kegel, and T. L. Cullen. (1964). Radiogeology of some high-background areas of Brazil. In: *The Natural Radiation Environment. Proceedings of an International Symposium*, Houston, TX, April 10–13, 1963 (Eds J. A. S. Adams and W. M. Lowder), William Marsh Rice University by The University of Chicago Press, 855–872.

Röttger, A., A. Honig, and D. Arnold. (2009). The German thoron progeny chamber – Concept and application. *Applied Radiation and Isotopes*, 67, 839–842. 10.1016/j.apadiso.2009.01.043.

Royal Society. (2001). *The Health Hazards of Depleted Uranium Munitions: Part I.* London, UK: The Royal Society Working Group on Health Hazards of Depleted Uranium Munitions.

Rubia, A. (2006). *LENA: Low Energy Neutrino Astronomy and Nucleon Decay Searches with Next Generation Large Underground Detectors*. 3rd ILIAS Annual Meeting, LNGS, March 2006, http://ilias.in2p3.fr/ilias_site/meetings/documents/FP7_030306/FP7_Rubbia.pdf.

Rutherford, E. (1900). A radio-active substance emitted from thorium compounds. *Philosophical Magazine*, 49, 1–14.

Rutherford, E. (1904). Disintegration of radioactive elements. *Harper's Monthly Magazine*, 108, 279–284.

Rutherford, E. (1905). The succession of changes in radioactive bodies. *Philosophical Transactions of the Royal Society of London, Series A*, 204, 169–219.

Rutherford, E., and H. L. Cooke. (1903). A Penetrating Radiation from the Earth's Surface. *Physical Review*, 16, 183.

Rutherford, P. M., M. J. Dudas, and J. M. Arocena. (1995a). Radium in phosphogypsum leachates. *Journal of Environmental Quality*, 24, 214–307.

Rutherford, P. M., M. J. Dudas, and J. M. Arocena. (1995b). Radon emanation coefficients for phosphogypsum. *Health Physics*, 69, 513–520.

Rutherford, P. M., M. J. Dudas, and J. M. Arocena. (1996). Heterogeneous distribution of radionuclides, barium and strontium in phosphogypsum by-product. *Science of the Total Environment*, 180, 201–209. 10.1016/0048-9697(95)04939-8.

Rutherford, P. M., M. J. Dudas, and R. A. Samek. (1994). Environmental impacts of phosphogypsum. *The Science for the Total Environment*, 149, 1–38.

Rutherford, E., and F. Soddy. (1902). The radioactivity of thorium compounds. *Journal of the Chemical Society Transactions*, 81, 837–860.

Ryan, J. M., and the Milagro Collaboration. (2005). The Forbush decreases of October-November 2003 as measured with *Milagro*. *Proceedings of the 29th International Cosmic Ray Conference*, ICRC, Pune, India, August 03–10, 2005. Vol. 1, pp. 383–386.

Saez Vergara, J.C., A. M. Romero Gutierrez, R. Rodríguez Jiminenz, R. D. Domínguez-Mompell Roman, P. Ortiz Garcia, and F. Merelo de Barbera. (2005). Monitoring of cosmic radiation on IBERIA comemcial flights: One year's experience of in-flight measurements. In: *The Natural Radiation Environment VII. Proceedings of an International Symposium*, Rhodes, Greece, 20–24 May 2002 (Eds J. P. McLaughlin, S. E. Simopoulus, and F. Steinhäusler), Elsevier, Amsterdam, pp. 885–893.

Saito, K., and P. Jacob. (1995). Gamma ray fields in the air due to sources in the ground. *Radiation Protection Dosimetry*, 58, 29–45.

Saito, K., N. Petoussi-Henss, and M. Zankl. (1998). Calculation of the effective dose and its variation from environmental gamma ray sources. *Health Physics*, 74, 698–706.

Samuelson, C. (1988). Retrospective determination of radon in houses. *Nature*, 334, 338.

Sappok, M., U. Quade, and T. Kluth. (1999). Melting of NORM – contaminated equipment of an offshore oil platform. *Waste Management 1999 Conference (WM'99), Proceedings of an International Conference*, Tucson, AZ, February 28–March 4.

Schery, S. D., and M. A. Wasiolek. (1998). Modelling radon flux from the Earth's surface In: *Radon and Thoron in the Human Environment. Proceedings of an International Symposium* (Eds A. Katase, and M. Shimo), World Scientific, Singapore, pp. 207–217.

Schmalz, R. F. (1990). Indoor radon in the United States and Canada – 1950–1984. In: *Environmental Radon: Occurrence, Control and Health Hazards* (Eds S. K. Manjumdar, R. F. Schmalz, and E. Willard Miller). The Pennsylvania Academy Press, Easton, PA, pp. 13–25.

Schofield, D. I., J. A. Evans, I. L. Millar, P. R. Wilby, and J. A. Aspden. (2008). New U–Pb and Rb–Sr constrains on pre-Acadian tectonism in North Wales. *Journal of the Geological Society*, 165(Part 5), 891–894.

Schötzig, U., and K. Debertin. (1983). Photon emission probabilities per decay of ^{226}Ra and ^{232}Th in equilibrium with their daughters products. *The International Journal of Applied Radiation and Isotopes*, 14, 533.

Schraube, H., G. Leuthold, W. Heinrich, and S. Roesler. (2002). *EPCARD-Program Package for the Calculation of Aviation Route Doses.* GSF-Report 08/02, Neuherberg.

Schrewe, U. J. (2000). Global measurements of the radiation exposure of civil air crew from 1997 to 1999. *Radiation Protection Dosimetry*, 91, 347–364.

Seaborg, G. T. (1958). *The Transuranium Elements*, Yale University Press. New Haven, CT.

Segovia, NS., S. de la Cruz Reyna, M. Mena, E. Ramos, M. Monnin, and J. L. Seidl. (1989). Radon-in-soil anomaly observed at Los Azufres geothermal field, Michoacán: A possible precursor of t he ($M_S = 8.1$) Mexico earthquake. *Natural Hazards*, 1, 319–329.

Seitz, R. (1996). Comparative assessment modeling. In: *Toxic Waste. Proceedings of the GTDC International Workshop*, Vienna, Global Technology Development Centre, pp. 178–199.

Sengupta, D., A. K. Mohanty, S. K. Das, and S. K. Saha. (2005). Natural radioactivity and radiation exposure in the high background area of the Chhatrapur beach placer deposits of Orissa, India. In: *The Natural Radiation Environment VII. Proceedings of an International Symposium*, Rhodes, Greece, 20–24 May 2002 (Eds J. P. McLaughlin, S. E. Simopoulus, and F. Steinhäusler), Elsevier, Amsterdam, pp. 1148–1151.

Shannon, L. V., and R. D. Cherry. (1971). Radium-226 in marine phytoplankton. *Earth Planetary Science Letters*, 11, 339–343.

Sheppard, C. W. (1944). Radioactivity and petroleum genesis. American Association of Petroleu Geologists (AAPG) Bulletin, 28, doi:10.1306/30933692-1681-1107-8645000102C11865D.

Shrock, R. R. (1982). *Geology at M.I.T., 1865–1965: A History of the First Hundred Years of Geology at the Massachusetts Institute of Technology,* Vol II Department of Operations and Projects. MIT Press, The Murray Printing Company, Westford, MA, 1102pp.

Shukla, V. K., T. V. Ramachandran, S. Chinnaesakki, S. J. Sartandel, and A. A. Shanbhag. (2005). Radiological impact of utilization of phosphogypsum and fly ash in building construction in India. In: *High Levels of Natural Radiation and Radon Areas: Radiation Dose and Health effects. Proceedings of an International Symposium,* International Congress Series, 1276, pp. 339–340.

Silva, N. C., E. A. de Nadai Fernandes, M. Cipriani, and M. H. T. Taddei. (2002). Leaching assessment of radioactive and non-radioactive elements from Brazilian phosphogypsum. In: *Technologically Enhanced Natural Radiation (TENR II). Proceedings of an International Symposium*, Rio de Janeiro, Brazil, 12–17 September 1999. International Atomic Energy Agency – IAEA-TECDOC-1271, Vienna, pp. 142–148.

Smith, A. R., H. R. Bowman, D. F. Mosier, F. Asaro, H. A. Wollenberg, and C.-Y. King. (1976). Investigation of Radon-222 in subsurface waters as an earthquake predictor. *IEEE Transactions on Nuclear Science*, 23, 694–698. 10.1109/TNS.1976.4326329.

Smits, F., and W. Gentner. (1950). Argon determination in potassium minerals I. Determination in tertiary potassium salts. In: *Geochronology: Radiometric Dating of Rocks and Minerals* (Ed. C. T. Harper), Douden, Hutchingson & Ross, Inc., Stroudsburg, PA (1973) pp. 312–318 – Translated from Argonbestimmungen and Kalium-Mineralien I. Bestimmungen an tertiären Kalisalzen. *Gerochim. Cosmichim. Acta*, I, pp. 22–27.

SNO Collaboration. (2004). Electron antineutrino search at the Sudbury Neutrino Observatory. *Physical Review D*, 70, 1–8.

Sohabi, M., and A. R. Esmaili. (2002). New public dose assessment of elevated natural radiation areas of Ramsar (Iran) for epidemiological studies. In: *High Levels of Natural Radiation and Radon Areas: Radiation Dose and Health Effects. Proceedings of an International Conference*, Munich, Germany, September 4–7, 2000, pp. 15–24.

Sorimachi, A., S. K. Sahoo, and S. Tokonami. (2009). Generation and control of thoron emanated from lantern mantles. *Review of Scientific Instruments*, 80, 015104–015104-4. 10.1063/1.3039413.

Sorimachi, A., Tokonami, S., Takahashi, H., and Kobayashi, Y. (2008). Performance of NIRS thoron chamber system. In: *The Natural Radiation Environment VIII. Proceedings of an International Symposium*, Búzios, Rio de Janeiro, Brazil, 7–12 October 2007 (Eds A. S. Paschoa and F. Steinhäusler), American Institute of Physics Conference Proceedings 1034, New York, pp. 206–209.

Spears, D. A., and M. R. Martinez-Tarrazona. (2004). Trace elements in combustion residues. *Fuel*, 83, 2265–2270. 10.1016/j.fuel.2004.06.025.

Spurny, F., and T. Daschev. (2002). On board aircrew dosimetry with a semiconductor spectrometer. *Radiation Protection Dosimetry*, 100, 525–528.

Spurny, F., I. Kovar, A. Malusek, and K. Turek. (2001). To the air crew exposure to cosmic radiation. *Proceedings of IRPA-10, P-1a-44.* Available in CD-ROM.

Spurny, F., K. Turek, B. Vicek, and T. Daschev. (2004). Aircrew exposure monitoring: Results of 2001 to 2003 studies. *Radiation Protection Dosimetry*, 110, 351–355.

Steinhäusler, F. (1975). Long-term measurements of Rn 222, Rn 220, Pb 214 and Pb 212 concentrations in the air of private and public buildings and their dependence on meteorological parameters. *Health Physics*, 29, 705–713.

Steinhäusler, F. (1985). Radiological impact on man and the environment from the oil and gas industry: Risk assessment for the critical group. In: *Radiation Safety Problems in the Caspian Region* (Eds M. K. Zaidi and I. Mustafaev). Springer, Netherlands, pp. 129–134. 10.1007/1-4020-2378-2_19.

Steinhäusler, F., and F. Edwards. (2005). *Catastrophic Terrorism and First Responders: Threats and Mitigation, NATO and Terrorism.* Springer, ISBN-10 1-4020-3583-7.

Steinhäusler, F., and A. S. Paschoa. (2005). The impact on man and the environment from the military use of depleted uranium: Worst case-scenario. In: *The Natural Radiation Environment VII. Proceedings of an International Symposium*, Rhodes, Greece, 20–24 May 2002 (Eds J. P. McLaughlin, S. E. Simopoulus, and F. Steinhäusler), Elsevier, Amsterdam, pp. 1047–1055.

Steinhäusler, F., S. Rydell, and L. Zaytseva. (2008). Risk due to radiological terror attacks with natural radionuclides. In: *The Natural Radiation Environment VIII. Proceedings of an International Symposium*, Búzios, Rio de Janeiro, Brazil, 7–12 October 2007 (Eds A. S. Paschoa and F. Steinhäusler), American Institute of Physics Conference Proceedings 1034, New York, pp. 3–15.

Steinhäusler, F., and L. Zaitseva. (2007). *Database on Nuclear Smuggling, Theft and Orphan Radiation Sources (DSTO).* Division of Physics, University of Salzburg, Salzburg, Austria (restricted access).

Stuiver, M., and P. D. Quay. (1980). Changes in atmospheric carbon-14 attributed to a variable sun. *Science*, 207, 11–19.

Suess, H. E. (1953). Natural radiocarbon and the rate of exchange of carbon dioxide between the atmosphere and the sea. *Nuclear Processes in Geologic Settings.* National Academy of Sciences, National Research Council Publication, Washington, D.C., pp. 52–56.

Sun, K., Q. Guo, and W. Zhuo. (2004). Feasibility of mapping radon exhalation rate from soil in China. *Journal of Nuclear Science and Technology*, 41, 86–90.

Swedjemark, G. A. (2004). The history of radon from a Swedish perspective. *Radiation Protection Dosimetry*, 109, 421–426. 101093/rpd/nch318.

Swedjemark, G. A., H. Mellander, and L. Mjönes. (1993). Radon levels in the 1988 Swedish Housing Stock. In: *Indoor Air Quality and Climate (Indoor Air'93). Proceedings an International Conference*, Helsinki, Finland, July 4–8, Vol. 4, pp. 491–496.

Swordy, S. (2001). The energy spectra and anisotropies of cosmic rays. *Space Science Reviews*, 99, 85–94.

Szerbin, P., L. Juhasz, I. Csige, A, Varhegi, J. Vincze, T. Szabo, and F.-J.-Maringer. (2005). Remediation case study of a coal fired power plant tailings pond. In: *Radioactivity in the Environment. Radioactivity in the Environment* (Eds J. P. McLaughlin, S. E. Simopoulos, F. Steinhäusler), Elsevier (M. S. Baxter, Series Editor), ISSN 1569-4860/doi:10.1016/S1569-4860(04)07110-4), Vol. 7, pp. 1071–1080.

Taber, A., and E. Normand. (1996). Single event upset in avionics. *IEEE Transactions on Nuclear Science*, NS-40(2), 2742–2750.

Tanner, A. B. (1964). Radon migration in the ground: A review. In: *The Natural Radiation Environment. Proceedings of an International Symposium*, Houston, TX, April 10–13, 1963 (Eds J. A. S. Adams and W. M. Lowder), William Marsh Rice University by The University of Chicago Press, pp. 161–190.

Tanner, A. B. (1980). Radon migration in the ground: A supplementary review. In: The Natural Radiation Environment III. In: *Proceedings of an International Symposium* (Eds T. F. Gesell and W. M. Lowder), CONF-780422, Springfield, VA, pp. 5–56.

Tanner, A.B. and C.R. Olsen. (1992). Technical Highlights of the Availability and Entry of the U.S. Department of Energy's Radon Research Program, Radiation Protection Dosimetry, 45, 367–372.

Tauhata, L., M. E. C. Vianna, A. E. de Oliveira, A. C. M. Ferreira, M. J. C. Bragança, A. F. Clain, and R. Q. de Faria. (2006). The Brazilian National Intercomparison Program (PNI/IRD/CNEN): Evaluation of 15 years of data. *Journal of Environmental Radioactivity*, 86, 384–390.

Tauhata, L., M. E. C. Vianna, A. E. de Oliveira, A. C. M. Ferreira, and C. C. S. da Conceição. (2002). Metrological capability of the Brazilian laboratories of analyses of radionuclides in environmental samples. *Applied Radiatation and Isotopes*, 56, 409–414.

Taylor, L. S. (1958). History of the International Commission on Radiological Units and Measurements (ICRU). *Health Physics*, 1, 306–307.

Taylor, G. C., R. D. Bentley, N. A. Horwood, R. Hunter, R. H. Iles, J. B. L. Jones, D. Powell, and D. J. Thomas. (2004). TEPC measurements in commercial aircraft. *Radiation Protection Dosimetry*, 110, 381–386.

Templeton, W. L. (1981). Dumping of low level radioactive waste in the deep ocean. In: *Impacts of Radionuclide Releases into the Marine Environment. Proceedings of an International Symposium*, IAEA, Vienna, pp. 451–464.

Templeton, W. L., R. E. Nakatani, and E. E. Held. (1971). Radiation effects. In; *Radioactivity in the Marine Environment*. National Academy of Sciences, National Research Council, Washington, D.C., pp. 6–11.

Terril, J. G., Jr., E. D. Harward, and I. P. Legget, Jr. (1967). Environmental aspects of nuclear and conventional power plants. *Industrial Medicine Surgery*, 36, 412–419.

The Borexino Collaboration. (2006). Search for electron antineutrino with the Borexino Counting Test Facility at Gran Sasso. *The European Physical Journal C – Particles and Fields*, 47, 21–30.

The Bratislava Report. (2009). Preliminary Findings and Recommendations (*The Bratislava Report*), NATO-Russia Advanced Research Workshop "Social and Psychological Effects of Radiological Terrorism", Radiological Terrorism: Public Response and Search for Resilience, Bratislava, Slovak Republic, Center for International Trade and Security, University of Georgia, August 31, 2009.

Tissot, B. P., and D. H. Welte. (1978). *Petroleum formation and occurrence: A new approach to oil and gas exploration*. Springer-Verlag, Berlin, ISBN-13: 9780387132815.

Tobias, C. A., and P. Todd. (1974). *Space Radiation Biology and Related Topics*. Academic Press, New York, 662 pp. ISBN 13: 9780126918502.

Tokonami, S., T. Ishikawa, A. Sorimachi, H. Takahashi, and N. Miyahara. (2008b). The Japanese radon and thoron reference chambers. In: *The Natural Radiation Environment VIII. Proceedings of an International Symposium*, Búzios, Rio de Janeiro, Brazil, 7–12 October 2007 (Eds A. S. Paschoa and F. Steinhäusler), American Institute of Physics Conference Proceedings 1034, New York, pp. 202–205.

Tokonami, S., T. Kovacs, M. Sugino, N. Kavasi, H. Takashi, Y. Kobayashi, A. Sorimachi, T. Ishikawa, and S. Yoshinaga. (2008a). Influence of environmental thoron on radon measurements and related issues. In: *The Natural Radiation Environment VIII. Proceedings of an International Symposium*, Búzios, Rio de Janeiro, Brazil, 7–12 October 2007 (Eds A. S. Paschoa and F. Steinhäusler), American Institute of Physics Conference Proceedings 1034, New York, pp. 145–148.

Tommasino, L., A. S. Paschoa, and M. C. Tommasino. (2008). *The Natural Radiation Environment VIII. Proceedings of an International Symposium*, Búzios, Rio de Janeiro, Brazil, 7–12 October 2007 (Eds A. S. Paschoa and F. Steinhäusler), American Institute of Physics Conference Proceedings 1034, New York, pp. 34–42.

Tomovic, R., and M. Vukobratovic. (1972). *General Sensitivity Theory*. Elsevier, New York, NY.

Trotti, F., S. Bucci, B. Dalzocchio, C. Zampieri, M. Lanciai, C. Innocenti, S. Maggiolo, L. Gaidolfi, and M. Belli. (2005). Towards the identification of work activities involving NORM in Italy. In: *Radioactivity in the Environment* (Eds J. P. McLaughlin, S. E. Simopoulos, F. Steinhäusler), Elsevier, (M. S. Baxter, Series Editor), ISSN 1569-4860/doi:10.1016/S1569-4860(04)07110-4), Vol. 7, pp. 973–984.

Trotti, F., C. Zampieri, E. Caldognetto, R. Ocone, A. Di Lullo, L. Magro, G. Jia, and G. Torri. (2008). A study concerning NORM in integrated steelworks. In: *Naturally Occurring Radioactive Material (NORM V). Proceedings of an International Symposium*, Seville, Spain, 19–22 March 2007. International Atomic Energy Agency, Vienna, pp. 351–354.

Tsai, K.-F., Y.-S. Lee, and H.-E. Chao. (2003). The study for recycling of NORM-contaminated steel scraps from steel industry. In: *Waste Management 2003 Conference (WM'03), Proceedings of an International Conference*, Tucson, AZ, February 23–27, 2003.

Turner, A. L. (1940). The nuclear fission. *Review of Modern Physics*, 12, 1–79.

Tveten, U., T. Haldorsen, and J. Reitan. (2000). Cosmic radiation and airline pilots: Exposure pattern as a function of aircraft type. *Radiation Protection Dosimetry*, 87, 157–165.

Tverdova, R. A., V. V. Fedina. (1974). Uranium in crudes of the lower Volga. *Geochemistry International*, 722 (translated into English from *Geokhimiya*, 7, 1045–1049).

Ulomov, V. I., and B. Z. Mavashev. (1971). Forerunner of Tashkent earthquakes. *Izvestiya Akademii Nauk Uzbekistana* USSR, pp. 188–200.

United Nations Environment Programme (UNEP). (2001). *Depleted Uranium in Kosovo: Post-Conflict Environmental Assessment*. UNEP, Switzerland, 186pp.

United Nations Environment Programme (UNEP). (2002). *Depleted Uranium in Serbia and Montenegro: Post-Conflict Environmental Assessment in the Federal Republic of Yugoslavia*. UNEP, Switzerland, 199pp.

United Nations Office at Geneva – UNOG. (2001). Convention on the Prohibition or Restriction on the Use of Certain Conventional Weapons which may be Deemed to be Excessively Injurious or to have Indiscriminate Effects as Amended on December 2001 (CCW) – also known as Inhumane Weapons Convention (CCW) – UNOG, Geneva, 7 pp.

United Nations Scientific Committee on the Effects of Atomic Radiation (UNSCEAR). (1956). Report to the General Assembly, United Nations, New York.

United Nations Scientific Committee on the Effects of Atomic Radiation (UNSCEAR). (1958). Report to the General Assembly, United Nations, New York.

United Nations Scientific Committee on the Effects of Atomic Radiation (UNSCEAR). (1977). Report to the General Assembly, with Annexes. Sources and Effects of Ionizing Radiation, United Nation, New York, 725pp.

United Nations Scientific Committee on the Effects of Atomic Radiation (UNSCEAR). (1988). Report to General Assembly. Sources, Effects and Risks of Ionizing Radiation. Report to the General Assembly, with Scientific Annexes. United Nations Sales Publication E.88.IX.2, United Nations, New York.

United Nations Scientific Committee on the Effects of Atomic Radiation (UNSCEAR). (1993). Report to General Assembly. Sources and Effects of Ionizing Radiation, Report to the General Assembly, with Scientific Annexes. Volume I: Sources, United Nations Publication Sales No. E.94.IX.2, United Nations, New York.

United Nations Scientific Committee on the Effects of Atomic Radiation (UNSCEAR). (2000). Report to the General Assembly, with Annexes. Sources and Effects of Ionizing Radiation, Vol. I; Sources, United Nation, New York, 654pp.

United States Department of Energy (US DoE). (2001). US Department of Energy Office of Environmental Management – *Depleted Uranium Hexafluoride Management Program – Depleted Uranium Fact Sheet*, Washington, D.C., 7pp.

United States Environmental Protection Agency (USEPA). (1993). *Diffuse NORM wastes – Waste characterization and preliminary assessment.* Office of Radiation and Indoor Air, April 1993.

United States Environmental Protection Agency (USEPA) (1994). Federal Radiation Protection Guidance for Exposure of the General Public. Notice, Part III, December 23, 1994. *Federal Register* 59, 66413–66428.

United States Geological Survey (USGS) (2003). Mineral commodity summaries, USGS, Reston, VA, Jnanuary 2003, http://minerals.usgs.gov/minerals/pubs/commodity/rare_earths/740303.pdf. Last accessed on 16 February 2009.

United States Geological Survey (USGS) (2008). Mineral commodity summaries, USGS, Reston, VA, January 2008, http://minerals.usgs.gov/minerals/pubs/commodity/gold/mcs-2008-gold.pdf. Last accessed on 5 January 2009.

Usoskin, I. G. (2008). A history of solar activity over millennia. *Living Reviews in Solar Physics*, 5, 3–87.

Usoskin, I. G., S. K. Solanki, and G. A. Kovaltsov. (2007). Grand minima and maxima of solar activity new observational constraints. *Astronomy and Astrophysics*, 471, 301–309.

US Patent. (1971). *Armour-Piercing Ammunition.* Patent number 3561363.

Uyttenhove, J. (2005). Capabilities and limitations of high-resolution gamma spectrometry in environmental research. In: *The Natural Radiation Environment VII. Proceedings of an International Symposium*, Rhodes, Greece, 20–24 May 2002 (Eds J. P. McLaughlin, S. E. Simopoulus, and F. Steinhäusler), Elsevier, Amsterdam, pp. 45–55.

Uyttenhove, J., M. Lemmens, and M. Zizi. (2002). Depleted uranium in Kosovo: Results of a survey by gamma spectrometry on soil samples. *Health Physics*, 83, 543–548.

Van Allen, J. A., and L. A. Frank. (1959). Radiation around the Earth to a radial distance of 107,400 km. *Nature*, 183, 430.

Van Allen, J. A., G. H. Ludwig, E. C. Ray, and C. E. McIlwain. (1958). Observations of high intensity radiation by satellites 1958 Alpha and Gamma. *Jet Propulsion*, 28, 588–592.

van der Heijde, H. B., P. J. Klijn, and W. F. Passchier. (1988). Radiological impacts of the disposal of phosphogypsum. *Radiation Protection Dosimetry*, 24, 419–423.

van der Pal, M., van der Spoel, de Meijer, R. J., Hendriks, and van der Graaf. (2005). Measurements and modeling of combined diffusive and advective radon transport in porous building materials. In: *The Natural Radiation Environment VII. Proceedings of an International Symposium*, Rhodes, Greece, 20–24 May 2002 (Eds J. P. McLaughlin, S. E. Simopoulus, and F. Steinhäusler), Elsevier, Amsterdam, pp. 371–380.

van Ginneken, A. (1975). *CASIM, Program to Simulate Hadronic Cascades in Bulk Matter.* Fermilab Report FN-272.

Vassiliou, A. H. (1980). The form of occurrence of uranium in deposits associated with organic matter. *Economic Geology*, 75, 609–617.

Verhaegen, F., and A. Poffijn. (2000). Air crew exposure on long-haul flights of the Belgian airlines. *Radiation Protection Dosimetry*, 88, 143–148.

Vernadsky, V. I. (1930). Sur les eaux naturelles riches en radium. *Comptes Rendus*, 190, 1172–1175.

Vernadsky, V. I., and W. Chlopin. (1932). Radium und mesothoriumhaltige natürliche Gewässer. *Zeitschrift fur Elektrochemie*, 8, 527–529.

Vianna, M. E. C. M., L. Tauhata, A. E. de Oliveira, J. P. de Oliveira, A. F. Clain, and A. C. M. Ferreira. (1998). Evaluation of Brazilian intercomparison program data from 1991 to 1995 of radionuclide assays in environmental samples. *Applied Radiation and Isotopes*, 49, 1463–1466.

von Baeckmann, A., G. Dillon, and D. Perricos. (1995). Nuclear verification in South Africa. *IAEA Bulletin*, 37/1(1), 42–48.

von Weizsäcker, C. F. (1937). The possibility of a dual beta decay of potassium. In: *Geochronology: Radiometric Dating of Rockes and Minerals* (Ed. C. T. Harper), Douden, Hutchingson & Ross, Inc., Stroudsburg, PA (1973), pp. 306–309 (translated from Physikalische Zeitschrift, 38, 623–624.

Walia, V., H. S. Virk, and B. S. Bajwa. (2006). Radon precursory signals for some earthquakes of magnitude >5 occurred in N-W Himalaya: An overview. *Pure and Applied Geophysics*, 163, 711–721. 10.1007/s00024-006-0044-z.

Walia, V., H. S. Virk, T. F. Yang, S. Mahajan, M. Walia, and B. S. Bajwa. (2005). Earthquake prediction studies using radon as a precursor in N-W Himalayas, India: A case study. *Terrestrial Atmospheric Oceanic Sciences*, 16, 775–804.

Wang. Z. (2002). Natural radiation environment in China. In: *The Natural Radiation Environment VII. Proceedings of an International Symposium*, Rhodes, Greece, 20–24 May 2002 (Eds J. P. McLaughlin, S. E. Simopoulus, and F. Steinhäusler), Elsevier, Amsterdam.

Ward, O. (1994). Deadly dose of radiation is murder … Russian-style, fears on rise about nuclear terrorism. *The Toronto Star*, 25 January, p. A18 (DSTO).

Washington, J. W., and A. W. Rose. (1992). Temporal variability of radon concentration in the interstitial gas of soils in Pennsylvania. *Journal of Geophysical Research*, 97B, 9145–9159.

Wasserburg, G. J., and R. J. Hayden. (1955). A^{40}–K^{40} dating. In: *Geochronology: Radiometric Dating of Rockes and Minerals* (Ed. C. T. Harper), Douden, Hutchingson & Ross, Inc., Stroudsburg, PA (1973), pp. 319–328, published originally in *Geochimica et Cosmochimica Acta*, 7, 51–60.

Wedepohl, K. H. (Ed.) (1978). *Handbook of Geochemistry*, Vol. II/5. Elements La(57) to U(92). 92-K-1 and 92-K-2, Springer Verlag, Berlin.

Wetherhill, G. W., L. T. Aldrich, and G. L. Davis. (1955). In: *Geochronology: Radiometric Dating of Rockes and Minerals* (Ed. C. T. Harper), Douden, Hutchingson & Ross, Inc., Stroudsburg, PA (1973), pp. 329–330, published originally in *Geochimica et Cosmochimica Acta*, 8, 171–172.

Whitcomb, R. (2007). Primary Spokesperson, Centres for Disease Control and Prevention (CDC). *Health Physics News*, XXXV(3), , March 2007.

Whitehead, W. L. (1954). Hydrocarbons formed by the effects of radioactivity and their role in the origin of petroleum. In: *Nuclear Geology, A Symposium on Nuclear Phenomena in the Earth Sciences* (Ed. H. Faul). Wiley, New York.

Wicke, A., and J. Postendörfer. (1982). Exposure to radon daughters in dwellings. In: *The Assessment of Radon and Daughter Exposure and Related Biological Effects. Proceedings of an International Seminar* (Eds G. F. Clemente, A. V. Nero, F. Steinhäusler, and M. E. Wrenn), RD Press, Salt Lake City, UT, pp. 135–143.

Wilkening, M.H., D. Stanley, and W.E. Clements. (1972). Radon-222 flux measurements in widely seperated regions: Rice University, Department of geology, Annual Report, U.S. Army of Engineers Water Experiment Station.

Wilkening, M. (1990). Radon in the environment. In: *Studies in Environmental Science*. Elsevier, Amsterdam, Vol. 40, 137pp.

Wilkinson, D. C., S. C. Daughtridge, J. L. Stone, H. H. Sauer, and P. Darling. (1991). TDRS-1 single event upsets and the effect of the space environment. *IEEE Transactions Nuclear Sciences*, NS-38(6), 1708–1712.

Wilson, C. T. R. (1901). On the ionisation of atmospheric air. *Proceedings of the Royal Society of London*, 68, 151–161.

Wilson, J. W. (1991). *Transport Methods and Interactions for Space Radiation*. NASA-1257.

Wilson, J. W. (1998). Overview of radiation environments and human exposures. In: *Proceedings of 34th Annual Meeting of the National Council on Radiation Protection and Measurements*, Washington, D.C., 1–2 April. Available electronically from http://techreprots.larc.nasa.gov/ltrs/ltrs.html

Wilson, J. W., L. W. Townsend, W. Schimmerling, G. S. Khandelwal, F. Khan, J. E. Nealy, F. A. Cucinotta, L. C. Simonsen, J. L. Shinn, J. W., and Norbury, J. W. (1999). Transport methods and interactions for space radiations, NASA Langley Research Center. Washington, D.C., National. Aeronautics Space Administration, (RP-1257).

Winde, F. and A. B. De Villiers, 2002. The nature and extent of uranium contamination from tailings dams in the Witwatersrand gold mining area (South Africa). In: . *Gold Tailings as a Source of Waterbourne Uranium Contamination of Streams – The Koekmoerspruit* (Eds F. Winde, P. Wade, and I. J. van der Walt, 2004), 2004 (Klerksdorp goldfield, South Africa) as a case study. Part I of III: Uranium Migration along the Aqueous Pathway, Water South Africa, Vol. 30, pp. 219–225. Available on website http://www.wrc.org.za.

Winde, F., P. Wade and I. J. Van der Walt. (2004). Gold tailings as a source of water-borne uranium contamination of streams – the Koekemoerspruit (South Africa) as a case study, Water SA, 30, 219–239.

Wollenberg, H. A., A. R. Smith, D. F. Mosier, S. Flexser, and M. Clark. (1984). Radon-222 in groundwater of the Long Valley caldera, California. *Pure and applied Geophysics*, 122, 327–339. 10.1007/BF00874602.

Woodhead, D. S. (1973). Levels of radioactivity in the marine environment and the dose commitment to marine organisms. In: *Radioactive Contamination of the Marine Environment. Proceedings of an International Symposium*, Seattle, Washington, D.C., 10–14 July 1972, IAEA, Vienna, pp. 499–525.

World Health Organization (WHO). (1959). *Effects of Radiation in Human Heredity: Investigations of Areas of High Natural Radiation*. Technical Report Series No. 166. Geneve, 47pp.

World Health Organization (WHO). (2001). Depleted Uranium – Sources, Exposure and Health effects, WHO/SDE/PHE/01.1, Geneva, 13pp.

World Health Organization (WHO). (1996). Working Group on Indoor Air Quality: A Risk-Based Approach to Health Criteria for Radon Indoors, Eilat, Israel, 28 March–4 April 1993. Regional Office for Europe, EUR/HFA target 21, 76 pp.

World Health Organization – WHO (2004). Radon and health, http://www.who.int?phe/radiation/en/2004Radon.

Wulf, Th. (1910). About the radiation of high penetration capacity contained in the atmosphere. *Physikalische Zeitschrift* [*Physics Magazine*], 10th Year, pp. 152–157.

Wymer, D. G. (2008). Managing Exposure to NORM – Consensus or Chaos? In: *Naturally Occurring Radioactive Material (NORM V). Proceedings of an International Symposium*, Seville, Span, 19–22 March 2007, International Atomic Energy Agency (IAEA), Vienna, pp. 31–56. ISBN 978-92-0-101508-2.

Yamada, J., M. Oka, M. Shimo, K. Minami, S. Minato, M. Sugimo, M. Hosoda, and M. Fukushi. (2008). The effect of geological and geographical features on environmental radiation. In: *The Natural Radiation Environment VIII. Proceedings of an International Symposium*, Búzios, Rio de Janeiro, Brazil, 7–12 October 2007 (Eds Anselmo S. Paschoa, and F. Steinhäusler), American Institute of Physics Conference Proceedings 1034, New York, pp. 498–502.

Yang, S., C. Li, and K. Yokoyama. (2006). Elemental composition and monazite age patterns of core sediments in the Changjiang Delta: Implications for sediment provenance and development history of the Changjiang River. *Earth and Planetary Science Letters*, 245, 762–776. 10.1016/j.epsl.2006.03.042.

Yang, S., H. Odah, and J. Shaw. (2000). Variations in the geomagnetic dipole moment over the last 12000 years. *Geophysical Journal International*, 140, 158–162.

Young, J. A., N. E. Erickson, and A. W. Fairhall. (1965). Tropospheric ^{14}C values in the Pacific Northwest and the Artic basin during 1964. In: *Radioactive Fallout from Nuclear Weapon Tests. Proceedings of an International Conference* (Ed. A. W. Klement, Jr.), Germantown, MD, November 3–4. US Atomic Energy Commission, CONF-765, pp. 422–427.

Ziegler, C. A. (1989). Technology and the process of scientific discovery: The case of cosmic rays. *Technology and Culture*, 30, 939–963. 10.2307/3106198.

Zobel, V., J. Eberth, U. Eberth, and E. Eube. (1977). ^{226}Ra as a calibration standard Ge(Li) spectrometers. *Nuclear Instruments and Methods*, 141, 329.

Zombeck, M. V. (2006). *Handbook of space astronomy and astrophysics*. Cambridge University Press, Cambridge, 3rd Ed., ISBN 0-521-78242-2.

Zuber, K. (2004). Neutrino Physics, Institute of Physics Series in High Energy Physics, Cosmology and gravitation, The Institute of Physics, London, 416pp. ISBN 07503 07501.

Author Index

Subject Index

FSC
www.fsc.org
MIX
Papier | Fördert gute Waldnutzung
FSC® C083411

Kolibri 360 GmbH
Ferdinand-Jühlke-Straße 7
99095 Erfurt, Deutschland
produktsicherheit@kolibri360.de